Semantic Web Technologies

Semantic Web Technologies Trends and Research in Ontology-based Systems

John Davies
BT, UK

Rudi Studer
University of Karlsruhe, Germany

Paul Warren
BT, UK

John Wiley & Sons, Ltd

Other Wiley Editorial Offices

John Wiley & Sons Inc., 111 River Street, Hoboken, NJ 07030, USA

Jossey-Bass, 989 Market Street, San Francisco, CA 94103-1741, USA

Wiley-VCH Verlag GmbH, Boschstr. 12, D-69469 Weinheim, Germany

John Wiley & Sons Australia Ltd, 42 McDougall Street, Milton, Queensland 4064, Australia

John Wiley & Sons (Asia) Pte Ltd, 2 Clementi Loop #02-01, Jin Xing Distripark, Singapore
129809

John Wiley & Sons Canada Ltd, 22 Worcester Road, Etobicoke, Ontario, Canada M9W 1L1

Library of Congress Cataloging-in-Publication Data

Davies, J. (N. John)
   Semantic Web technologies : trends and research in ontology-based systems
   / John Davies, Rudi Studer, Paul Warren.
      p. cm.
   Includes bibliographical references and index.
   ISBN-13: 978-0-470-02596-3 (cloth : alk. paper)
   ISBN-10: 0-470-02596-4 (cloth : alk. paper)
   1. Semantic Web.   I. Studer, Rudi.   II. Warren, Paul. III.   Title: Trends
and research in ontology-based systems.   IV. Title.

TK5105.88815.D38 2006
025.04–dc22   2006006501

British Library Cataloguing in Publication Data

A catalogue record for this book is available from the British Library

ISBN-13: 978-0-470-02596-3
ISBN-10: 0-470-02596-4

Typeset in 10/11.5 pt Palatino by Thomson Press (India) Ltd, New Delhi, India
Printed and bound in Great Britain by Antony Rowe Ltd, Chippenham, Wiltshire
This book is printed on acid-free paper responsibly manufactured from sustainable forestry
in which at least two trees are planted for each one used for paper production.

Contents

Foreword

Semantically Enabled Knowledge Technologies—Toward a New Kind of Web

Information technology has a surprising way of changing our culture radically—often in ways unimaginable to the inventors.

When Gutenberg developed moveable type in the middle of the fifteenth century, his primary goal was to develop a mechanism to speed the printing of Bibles. Gutenberg probably never thought of his technology in terms of the general dissemination of human knowledge via printed media. He never planned explicitly for printing presses to democratize the ownership of knowledge and to take away the monopoly on the control of information that had been held previously by the Church—which initially lacked Gutenberg's technology, but which had at its disposal the vast numbers of dedicated personnel needed to store, copy, and distribute books in a totally manual fashion. Gutenberg sought a better way to produce Bibles, and as a result changed fundamentally the control of knowledge in Western society. Within a few years, anyone who owned a printing press could distribute knowledge widely to anyone willing to read it.

In the late twentieth century, Berners-Lee had the goal of providing rapid, electronic access to the online technical reports and other documents created by the world's high-energy physics laboratories. He sought to make it easier for physicists to access their arcane, distributed literature from a range of research centers scattered about the world. In the process, Berners-Lee laid the foundation for the World Wide Web. In 1989, Berners-Lee could only begin imagine how his proposal to link technical reports via hypertext might someday change fundamentally essential aspects of human communication and social interaction. It was not his intention to revolutionize communication of information for e-commerce, for geographic reasoning, for government services, or for any of the myriad Web-based applications that we now take for granted.

Our society changed irreversibly, however, when Berners-Lee invented HTML and HTTP.

The World Wide Web provides a dazzling array of information services—designed for use by people—and has become an ingrained part of our lives. There is another Web coming, however, where online information will be accessed by intelligent agents that will be able to reason about that information and communicate their conclusions in ways that we can only begin to dream about. This Semantic Web represents the next stage in the evolution of communication of human knowledge. Like Gutenberg, the developers of this new technology have no way of envisioning the ultimate ramifications of their work. They are, however, united by the conviction that creating the ability to capture knowledge in machine understandable form, to publish that knowledge online, to develop agents that can integrate that knowledge and reason about it, and to communicate the results both to people and to other agents, will do nothing short of revolutionize the way people disseminate and utilize information.

The European Union has long maintained a vision for the advent of the "information society," supporting several large consortia of academic and industrial groups dedicated to the development of infrastructure for the Semantic Web. One of these consortia has had the goal of developing Semantically Enabled Knowledge Technologies (SEKT; http://www.sekt-project.com), bringing together fundamental research, work to build novel software components and tools, and demonstration projects that can serve as reference implementations for future developers.

The SEKT project has brought together some of Europe's leading contributors to the development of knowledge technologies, data-mining systems, and technologies for processing natural language. SEKT researchers have sought to lay the groundwork for scalable, semi-automatic tools for the creation of ontologies that capture the concepts and relationships among concepts that structure application domains; for the population of ontologies with content knowledge; and for the maintenance and evolution of these knowledge resources over time. The use of ontologies (and of procedural middleware and Web services that can operate on ontologies) emerges as the fundamental basis for creating intelligence on the Web, and provides a unifying framework for all the work produced by the SEKT investigators.

This volume presents a review and synopsis of current methods for engineering the Semantic Web while also documenting some of the early achievements of the SEKT project. The chapters of this book provide overviews not only of key aspects of Semantic Web technologies, but also of prototype applications that offer a glimpse of how the Semantic Web will begin to take form in practice. Thus, while many of the chapters deal with specific technologies such as those for Semantic Web services, metadata extraction, ontology alignment, and ontology engineering, the case studies provide examples of how these technologies can come together to solve real-world problems using Semantic Web techniques.

In recent years, many observers have begun to ask hard questions about what the Semantic Web community has achieved and what it can promise. The prospect of Web-based intelligence is so alluring that the scientific community justifiably is seeking clarity regarding the current state of the technology and what functionality is really on the horizon. In this regard, the work of the SEKT consortium provides an excellent perspective on contemporary research on Semantic Web infrastructure and applications. It also offers a glimpse of the kinds of knowledge-based resources that, in a few years time, we may begin to take for granted— just as we do current-generation text-based Web browsers and resources.

At this point, there is no way to discern whether the Semantic Web will affect our culture in a way that can ever begin to approximate the changes that have resulted from the invention of print media or of the World Wide Web as we currently know it. Indeed, there is no guarantee that many of the daunting problems facing Semantic Web researchers will be solved anytime soon. If there is anything of which we can be sure, however, it is that even the SEKT researchers cannot imagine all the ways in which future workers will tinker with Semantic Web technologies to engineer, access, manage, and reason with heterogeneous, distributed knowledge stores. Research on the Semantic Web is helping us to appreciate the enormous possibilities of amassing human knowledge online, and there is justifiable excitement and anticipation in thinking about what that achievement might mean someday for nearly every aspect of our society.

Mark A. Musen
Stanford, California, USA
January 2, 2006

1

Introduction

Paul Warren, Rudi Studer and John Davies

1.1. SEMANTIC WEB TECHNOLOGIES

That we need a new approach to managing information is beyond doubt. The technological developments of the last few decades, including the development of the World Wide Web, have provided each of us with access to far more information than we can comprehend or manage effectively. A Gartner study (Morello, 2005) found that 'the average knowledge worker in a Fortune 1000 company sends and receives 178 messages daily', whilst an academic study has shown that the volume of information in the public Web tripled between 2000 and 2003 (Lyman et al., 2005). We urgently need techniques to help us make sense of all this; to find what we need to know and filter out the rest; to extract and summarise what is important, and help us understand the relationships between it. Peter Drucker has pointed out that knowledge worker productivity is the biggest challenge facing organisations (Drucker, 1999). This is not surprising when we consider the increasing proportion of knowledge workers in the developing world. Knowledge management has been the focus of considerable attention in recent years, as comprehensively reviewed in (Holsapple, 2002). Tools which can significantly help knowledge workers achieve increased effectiveness will be tremendously valuable in the organisation.

At the same time, integration is a key challenge for IT managers. The costs of integration, both within an organisation and with external trading partners, are a significant component of the IT budget. Charlesworth (2005) points out that information integration is needed to 'reach a better understanding of the business through its data', that is to achieve a

Semantic Web Technologies: Trends and Research in Ontology-based Systems
John Davies, Rudi Studer, Paul Warren © 2006 John Wiley & Sons, Ltd

common view of all the data and understand their relationships. He describes application integration, on the other hand, as being concerned with sharing 'data, information and business and processing logic between disparate applications'. This is driven in part by the need to integrate new technology with legacy systems, and to integrate technology from different suppliers. It has given rise to the concept of the service oriented architecture (SOA), where business functions are provided as loosely coupled services. This approach provides for more flexible loose coupling of resources than in traditional system architecture, and encourages reuse. Web services are a natural, but not essential, way of implementing an SOA. In any case, the need is to identify and integrate the required services, whilst at the same time enabling the sharing of data between services.

For their effective implementation, information management, information integration and application integration all require that the underlying information and processes be described and managed semantically, that is they are associated with a machine-processable description of their meaning. This, the fundamental idea behind the Semantic Web became prominent at the very end of the 1990s (Berners-Lee, 1999) and in a more developed form in the early 2000s (Berners-Lee *et al.*, 2001). The last half decade has seen intense activity in developing these ideas, in particular under the auspices of the World Wide Web Consortium (W3C).[1] Whilst the W3C has developed the fundamental ideas and standardised the languages to support the Semantic Web, there has also been considerable research to develop and apply the necessary technologies, for example natural language processing, knowledge discovery and ontology management. This book describes the current state of the art in these technologies.

All this work is now coming to fruition in practical applications. The initial applications are not to be found on the global Web, but rather in the world of corporate intranets. Later chapters of this book describe a number of such applications.

The book was motivated by work carried out on the SEKT project (http://www.sekt-project.com). Many of the examples, including two of the applications, are drawn from this project. However, it is not biased towards any particular approach, but offers the reader an overview of the current state of the art across the world.

1.2. THE GOAL OF THE SEMANTIC WEB

The Semantic Web and Semantic Web technologies offer us a new approach to managing information and processes, the fundamental principle of which is the creation and use of semantic metadata.

---

[1] See: http://www.w3.org/2001/sw/

For information, metadata can exist at two levels. On the one hand, they may describe a document, for example a web page, or part of a document, for example a paragraph. On the other hand, they may describe entities within the document, for example a person or company. In any case, the important thing is that the metadata is semantic, that is it tells us about the content of a document (e.g. its subject matter, or relationship to other documents) or about an entity within the document. This contrasts with the metadata on today's Web, encoded in HTML, which purely describes the format in which the information should be presented: using HTML, you can specify that a given string should be displayed in bold, red font but you cannot specify that the string denotes a product price, or an author's name, and so on.

There are a number of additional services which this metadata can enable (Davies *et al.*, 2003).

In the first place, we can organise and find information based on meaning, not just text. Using semantics our systems can understand where words or phrases are equivalent. When searching for 'George W Bush' we may be provided with an equally valid document referring to 'The President of the U.S.A.'. Conversely they can distinguish where the same word is used with different meanings. When searching for references to 'Jaguar' in the context of the motor industry, the system can disregard references to big cats. When little can be found on the subject of a search, the system can try instead to locate information on a semantically related subject.

Using semantics we can improve the way information is presented. At its simplest, instead of a search providing a linear list of results, the results can be clustered by meaning. So that a search for 'Jaguar' can provide documents clustered according to whether they are about cars, big cats, or different subjects all together. However, we can go further than this by using semantics to merge information from all relevant documents, removing redundancy, and summarising where appropriate. Relationships between key entities in the documents can be represented, perhaps visually. Supporting all this is the ability to reason, that is to draw inferences from the existing knowledge to create new knowledge.

The use of semantic metadata is also crucial to integrating information from heterogeneous sources, whether within one organisation or across organisations. Typically, different schemas are used to describe and classify information, and different terminologies are used within the information. By creating mappings between, for example, the different schemas, it is possible to create a unified view and to achieve interoperability between the processes which use the information.

Semantic descriptions can also be applied to processes, for example represented as web services. When the function of a web service can be described semantically, then that web service can be discovered more easily. When existing web services are provided with metadata describing their function and context, then new web services can be automatically composed by the combination of these existing web services. The use of such semantic descriptions is likely to be essential to achieve large-scale implementations of an SOA.

1.3. ONTOLOGIES AND ONTOLOGY LANGUAGES

At the heart of all Semantic Web applications is the use of ontologies. A commonly agreed definition of an ontology is: 'An ontology is an explicit and formal specification of a conceptualisation of a domain of interest' (c.f. Gruber, 1993). This definition stresses two key points: that the conceptualisation is formal and hence permits reasoning by computer; and that a practical ontology is designed for some particular domain of interest. Ontologies consist of concepts (also knowns as classes), relations (properties), instances and axioms and hence a more succinct definition of an ontology is as a 4-tuple $\langle C, R, I, A \rangle$, where C is a set of concepts, R a set of relations, I a set of instances and A a set of axioms (Staab and Studer, 2004).

Early work in Europe and the US on defining ontologies languages has now converged under the aegis of the W3C, to produce a Web Ontology Language, OWL.[2]

The OWL language provides mechanisms for creating all the components of an ontology: concepts, instances, properties (or relations) and axioms. Two sorts of properties can be defined: object properties and datatype properties. Object properties relate instances to instances. Datatype properties relate instances to datatype values, for example text strings or numbers. Concepts can have super and subconcepts, thus providing a mechanism for subsumption reasoning and inheritance of properties. Finally, axioms are used to provide information about classes and properties, for example to specify the equivalence of two classes or the range of a property.

In fact, OWL comes in three species. OWL Lite offers a limited feature set, albeit adequate for many applications, but at the same time being relatively efficient computationally. OWL DL, a superset of OWL Lite, is based on a form of first order logic known as Description Logic. OWL Full, a superset of OWL DL, removes some restrictions from OWL DL but at the price of introducing problems of computational tractability. In practice much can be achieved with OWL Lite.

OWL builds on the Resource Description Framework (RDF)[3] which is essentially a data modelling language, also defined by the W3C. RDF is graph-based, but usually serialised as XML. Essentially, it consists of triples: subject, predicate, object. The subject is a resource (named by a

---

[2] See: http://www.w3.org/2004/OWL/
[3] See: http://www.w3.org/RDF/

URI), for example an instance, or a blank node (i.e., not identifiable outside the graph). The predicate is also a resource. The object may be a resource, blank node, or a Unicode string literal.

For a full introduction to the languages and basic technologies underlying the Semantic Web see [Antoniou and van Harmelen, 2004].

1.4. CREATING AND MANAGING ONTOLOGIES

The book is organized broadly to follow the lifecycle of an ontology, that is discussing technologies for ontology creation, management and use, and then looking in detail at some particular applications. This section and the two which follow provide an overview of the book's structure.

The construction of an ontology can be a time-consuming process, requiring the services of experts both in ontology engineering and the domain of interest. Whilst this may be acceptable in some high value applications, for widespread adoption some sort of semiautomatic approach to ontology construction will be required. Chapter 2 explains how this is possible through the use of knowledge discovery techniques.

If the generation of ontologies is time-consuming, even more is this the case for metadata extraction. Central to the vision of the Semantic Web, and indeed to that of the semantic intranet, is the ability to automatically extract metadata from large volumes of textual data, and to use this metadata to annotate the text. Chapter 3 explains how this is possible through the use of information extraction techniques based on natural language analysis.

Ontologies need to change, as knowledge changes and as usage changes. The evolution of ontologies is therefore of key importance. Chapter 4 describes two approaches, reflecting changing knowledge and changing usage. The emphasis is on evolving ontologies incrementally. For example, in a situation where new knowledge is continuously being made available, we do not wish to have to continuously recompute our ontology from scratch.

Reference has already been made to the importance of being able to reason over ontologies. Today an important research theme in machine reasoning is the ability to reason in the presence of inconsistencies. In classical logic any formula is a consequence of a contradiction, that is in the presence of a contradiction any statement can be proven true. Yet in the real world of the Semantic Web, or even the semantic intranet, inconsistencies will exist. The challenge, therefore, is to return meaningful answers to queries, despite the presence of inconsistencies. Chapter 5 describes how this is possible.

A commonly held misconception about the Semantic Web is that it depends on the creation of monolithic ontologies, requiring agreement from many parties. Nothing could be further from the truth. Of course, it is good design practice to reuse existing ontologies wherever possible, particularly where an ontology enjoys wide support. However, in many cases we need to construct mappings between ontologies describing the same domain, or alternatively merge ontologies to form their union. Both approaches rely on the identification of correspondences between the ontologies, a process known as ontology alignment, and one where (semi-)automatic techniques are needed. Chapter 6 describes techniques for ontology merging, mapping and alignment.

1.5. USING ONTOLOGIES

Chapter 7 explains two rather different roles for ontologies in knowledge management, and discusses the different sorts of ontologies: upper-level versus domain-specific; light-weight versus heavy weight. The chapter illustrates this discussion with reference to the PROTON ontology.[4]

Chapter 8 describes the state of the art in three aspects of ontology-based information access: searching and browsing; natural language generation from structured data, for example described using ontologies; and techniques for on-the-fly repurposing of data for a variety of devices. In each case the chapter discusses current approaches and their limitations, and describes how semantic web technology can offer an improved user experience. The chapter also describes a semantic search agent application which encompasses all three aspects.

The creation of ontologies, although partially automated, continues to require human intervention and a methodology for that intervention. Previous methodologies for introducing knowledge technologies into the organisation have tended to assume a centralised approach which is inconsistent with the flexible ways in which modern organisations operate. The need today is for a distributed evolution of ontologies. Typically individual users may create their own variations on a core ontology, which then needs to be kept in step to reflect the best of the changes introduced by users. Chapter 9 discusses the use of such a methodology.

Ontologies are being increasingly seen as a technology for streamlining the systems integration process, for example through the use of semantic descriptions for web services. Current web services support inter-operability through common standards, but still require considerable human interaction, for example to search for web services and then to combine them in a useful way. Semantic web services, described in Chapter 10, offer the possibility of automating web service discovery, composition and invocation. This will have considerable impact in areas such as e-Commerce and Enterprise Application Integration, by

---

[4] http://proton.semanticweb.org/

enabling dynamic and scalable cooperation between different systems and organizations.

1.6. APPLICATIONS

There are myriad applications for Semantic Web technology, and it is only possible in one book to cover a small fraction of them. The three described in this book relate to specific business domains or industry sectors. However, the general principles which they represent are relevant across a wide range of domains and sectors.

Chapter 11 describes the key role which Semantic Web technology is playing in enhancing the concept of a Digital Library. Interoperability between digital libraries is seen as a 'Grand Challenge', and Semantic Web technology is key to achieving such interoperability. At the same time, the technology offers new ways of classifying, finding and presenting knowledge, and also the interrelationships within a corpus of knowledge. Moreover, digital libraries are one example of intelligent content management systems, and much of what is discussed in Chapter 11 is applicable generally to such systems.

Chapter 12 looks at an application domain within a particular sector, the legal sector. Specifically, it describes how Semantic Web technology can be used to provide a decision support system for judges. The system provides the user with responses to natural language questions, at the same time as backing up these responses with reference to the appropriate statutes. Whilst apparently very specific, this can be extended to decision support in general. In particular, a key challenge is combining everyday knowledge, based on professional experience, with formal legal knowledge contained in statute databases. The development of the question and answer database, and of the professional knowledge ontology to describe it, provide interesting examples of the state of the art in knowledge elicitation and ontology development.

The final application, in Chapter 13, builds on the semantic web services technology in Chapter 10, to describe how this technology can be used to create an SOA. The approach makes use of the Web Services Modelling Ontology (WSMO)[5] and permits a move away from point to point integration which is costly and inflexible if carried out on a large scale. This is particularly necessary in the telecommunications industry, where operational support costs are high and customer satisfaction is a key differentiator. Indeed, the approach is valuable wherever IT systems need to be created and reconfigured rapidly to support new and rapidly changing customer services.

---

[5] See http://www.wsmo.org/

1.7. DEVELOPING THE SEMANTIC WEB

This book aims to provide the reader with an overview of the current state of the art in Semantic Web technologies, and their application. It is hoped that, armed with this understanding, readers will feel inspired to further develop semantic web technologies and to use semantic web applications, and indeed to create their own in their industry sectors and application domains. In this way they can achieve real benefit for their businesses and for their customers, and also participate in the development of the next stage of the Web.

REFERENCES

Antoniou G, van Harmelen F. 2004. *A Semantic Web Primer*. The MIT Press: Cambridge, Massachusetts.

Berners-Lee T. 1999. *Weaving the Web*. Orion Business Books.

Berners-Lee T, Hendler J, Lassila O. 2001. The semantic web. In *Scientific American*, May 2001.

Charlesworth I. 2005. Integration fundamentals, Ovum.

Davies J, Fensel D, van Harmelen F (eds). 2003. *Towards the Semantic Web: Ontology-Driven Knowledge Management*. John Wiley & Sons, Ltd. ISBN: 0470848677.

Drucker P. 1999. Knowledge worker productivity: the biggest challenge. *California Management Review* 41(2):79–94.

Fensel D, Hendler JA, Lieberman H, Wahlster W (eds). 2003. *Spinning the Semantic Web: Bringing the World Wide Web to its Full Potential*. MIT Press: Cambridge, MA. ISBN 0-262-06232-1.

Gruber T. 1993. A translation approach to portable ontologies. *Knowledge Acquisition* 5(2):199–220, http://ksl-web.stanford.edu/KSL_Abstracts/KSL-92-71.html

Holsapple CW Eds. 2002. *Handbook on Knowledge Management*. Springer: ISBN:3540435271.

Lyman P, *et al.* 2005. How Much Information? 2003, School of Information Management and Systems, University of California at Berkeley, http://www.sims.berkeley.edu/research/projects/how-much-info-2003/

Morello D. 2005. The human impact of business IT: How to Avoid Diminishing Returns.

Staab S, Studer R (Eds). 2004. *Handbook on Ontologies. International Handbooks on Information Systems*. Springer: ISBN 3-540-40834-7.

2

Knowledge Discovery for Ontology Construction

Marko Grobelnik and Dunja Mladenić

2.1. INTRODUCTION

We can observe that the focus of modern information systems is moving from 'data-processing' towards 'concept-processing', meaning that the basic unit of processing is less and less is the atomic piece of data and is becoming more a semantic concept which carries an interpretation and exists in a context with other concepts. As mentioned in the previous chapter, an ontology is a structure capturing semantic knowledge about a certain domain by describing relevant concepts and relations between them.

Knowledge Discovery (KD) is a research area developing techniques that enable computers to discover novel and interesting information from raw data. Usually the initial output from KD is further refined via an iterative process with a human in the loop in order to get knowledge out of the data. With the development of methods for semi-automatic processing of complex data it is becoming possible to extract hidden and useful pieces of knowledge which can be further used for different purpose including semi-automatic ontology construction. As ontologies are taking a significant role in the Semantic Web, we address the problem of semi-automatic ontology construction supported by Knowledge Discovery. This chapter presents several approaches from Knowledge Discovery that we envision as useful for the Semantic Web and in particular for semi-automatic ontology construction. In that light, we propose to decompose the semi-automatic ontology construction process

Semantic Web Technologies: Trends and Research in Ontology-based Systems
John Davies, Rudi Studer, Paul Warren © 2006 John Wiley & Sons, Ltd

into several phases. Several scenarios of the ontology learning phase are identified based on different assumptions regarding the provided input data. We outline some ideas how the defined scenarios can be addressed by different Knowledge Discovery approaches.

The rest of this Chapter is structured as follows. Section 2.2 provides a brief description of Knowledge Discovery. Section 2.3 gives a definition of the term ontology. Section 2.4 describes the problem of semi-automatic ontology construction. Section 2.5 describes the proposed methodology for semi-automatic ontology construction where the whole process is decomposed into several phases. Section 2.6 describes several Knowledge Discovery methods in the context of the semi-automatic ontology construction phases defined in Section 2.5. Section 2.7 gives a brief overview of the existing work in the area of semi-automatic ontology construction. Section 2.8 concludes the Chapter with discussion.

2.2. KNOWLEDGE DISCOVERY

The main goal of Knowledge Discovery is to find useful pieces of knowledge within the data with little or no human involvement. There are several definitions of Knowledge Discovery and here we cite just one of them: Knowledge Discovery is a process which aims at the extraction of interesting (nontrivial, implicit, previously unknown and potentially useful) information from data in large databases (Fayad et al., 1996).

In Knowledge Discovery there has been recently an increased interest for learning and discovery in unstructured and semi-structured domains such as text (Text Mining), web (Web Mining), graphs/networks (Link Analysis), learning models in relational/first-order form (Relational Data Mining), analyzing data streams (Stream Mining), etc. In these we see a great potential for addressing the task of semi-automatic ontology construction.

Knowledge Discovery can be seen as a research area closely connected to the following research areas: *Computational Learning Theory* with a focus on mainly theoretical questions about learnability, computability, design and analysis of learning algorithms; *Machine Learning* (Mitchell, 1997), where the main questions are how to perform automated learning on different kinds of data and especially with different representation languages for representing learned concepts; *Data-Mining* (Fayyad et al., 1996; Witten and Frank, 1999; Hand et al., 2001), being rather applied area with the main questions on how to use learning techniques on large-scale real-life data; *Statistics* and statistical learning (Hastie et al., 2001) contributing techniques for data analysis (Duda et al., 2000) in general.

2.3. ONTOLOGY DEFINITION

Ontologies are used for organizing knowledge in a structured way in many areas—from philosophy to Knowledge Management and the

Semantic Web. We usually refer to an ontology as a graph/network structure consisting from:

1. a set of concepts (vertices in a graph);
2. a set of relationships connecting concepts (directed edges in a graph);
3. a set of instances assigned to a particular concepts (data records assigned to concepts or relation).

More formally, an ontology is defined (Ehrig *et al.*, 2005) as a structure $O = (C, T, R, A, I, V, \leq_C, \leq_T, \sigma_R, \sigma_A, \iota_C, \iota_T, \iota_R, \iota_A)$. It consists of disjoint sets of concepts (*C*), types (*T*), relations (*R*), attributes (*A*), instances (*I*), and values (*V*). The partial orders $\leq_C$ (on *C*) and $\leq_T$ (on *T*) define a concept hierarchy and a type hierarchy, respectively. The function $\sigma_R: R \rightarrow C^2$ provides relation signatures (i.e., for each relation, the function specifies which concepts may be linked by this relation), while $\sigma_A: A \rightarrow C \times T$ provides attribute signatures (for each attribute, the function specifies to which concept the attribute belongs and what is its datatype). Finally, there are partial instantiation functions $\iota_C: C2^I$ (the assignment of instances to concepts), $\iota_T: T2^V$ (the assignment of values to types), $\iota_R: R \rightarrow 2^{I \times I}$ (which instances are related by a particular relation), and $\iota_A: A \rightarrow 2^{I \times V}$ (what is the value of each attribute for each instance). Another formalization of ontologies, based on similar principles, has been described by Bloehdorn *et al.* (2005). Notice that this theoretical framework can be used to define evaluation of ontologies as a function that maps the ontology O to a real number (Brank *et al.*, 2005).

2.4. METHODOLOGY FOR SEMI-AUTOMATIC ONTOLOGY CONSTRUCTION

Knowledge Discovery technologies can be used to support different phases and scenarios of semi-automatic ontology construction. We believe that today a completely automatic construction of good quality ontologies is in general not possible for theoretical, as well as practical reasons (e.g., the soft nature of the knowledge being conceptualized). As in Knowledge Discovery in general, human interventions are necessary but costly in terms of resources. Therefore the technology should help in efficient utilization of human interventions, providing suggestions, highlighting potentially interesting information, and enabling refinements of the constructed ontology.

There are several definitions of the ontology engineering and construction methodology, mainly based on a knowledge management perspective. For instance, the DILIGENT ontology engineering methodology described in Chapter 9 defines five main steps of ontology engineering: building, local adaptation, analysis, revision, and local update. Here, we define a methodology for *semi-automatic ontology construction* analogous to the CRISP-DM methodology (Chapman *et al.*, 2000) defined for the Knowledge Discovery process. CRISP-DM involves six interrelated phases: business understanding, data understanding, data preparation, modeling, evaluation, and deployment. From the perspective of Knowledge Discovery, semi-automatic ontology construction can be defined as consisting of the following interrelated phases:

1. *domain understanding* (what is the area we are dealing with?);
2. *data understanding* (what is the available data and its relation to semi-automatic ontology construction?);
3. *task definition* (based on the available data and its properties, define task(s) to be addressed);
4. *ontology learning* (semi-automated process addressing the task(s) defined in the phase 3);
5. *ontology evaluation* (estimate quality of the solutions to the addressed task(s)); and
6. *refinement with human in the loop* (perform any transformation needed to improve the ontology and return to any of the previous steps, as desired).

The first three phases require intensive involvement of the user and are prerequisites for the next three phases. While phases 4 and 5 can be automated to some extent, the last phase heavily relays on the user. Section 2.5 describes the fourth phase and some scenarios related to addressing the ontology learning problem by Knowledge Discovery methods. Using Knowledge Discovery in the fifth phase for semi-automatic ontology evaluation is not in the scope of this Chapter, an overview can be found in (Brank *et al.*, 2005).

2.5. ONTOLOGY LEARNING SCENARIOS

From a Knowledge Discovery perspective, we see an ontology as just another class of models (somewhat more complex compared to typical Machine Learning models) which needs to be expressed in some kind of hypothesis language. Depending on the different assumptions regarding the provided input data, ontology learning can be addressed via different tasks: learning just the ontology concepts, learning just the ontology relationships between the existing concepts, learning both the concepts and relations at the same time, populating an existing ontology/structure, dealing with dynamic data streams, simultaneous construction of ontologies giving different views on the same data, etc. More formally, we define the ontology learning tasks in terms of mappings between ontology components, where some of the components are given and some are missing and we want to induce the missing ones. Some typical scenarios in ontology learning are the following:

1. Inducing concepts/clustering of instances (given instances).
2. Inducing relations (given concepts and the associated instances).
3. Ontology population (given an ontology and relevant, but not associated instances).
4. Ontology generation (given instances and any other background information).
5. Ontology updating/extending (given an ontology and background information, such as, new instances or the ontology usage patterns).

Knowledge discovery methods can be used in all of the above typical scenarios of ontology learning. When performing the learning using Knowledge Discovery, we need to select a language for representation of a membership function. Examples of different representation languages as used by machine learning algorithms are: Linear functions (e.g., used by Support-Vector-Machines), Propositional logic (e.g., used in decision trees and decision rules), First order logic (e.g., used in Inductive Logic programming). The representation language selected informs the expressive power of the descriptions and complexity of computation.

2.6. USING KNOWLEDGE DISCOVERY FOR ONTOLOGY LEARNING

Knowledge Discovery techniques are in general aiming at discovering knowledge and that is often achieved by finding some structure in the data. This means that we can use these techniques to map unstructured data sources, such as a collection of text documents, into an ontological structure. Several techniques that we find relevant for ontology learning have been developed in Knowledge Discovery, some of them in combination with related fields such as Information Retrieval (van Rijsbergen, 1979) and Language Technologies (Manning and Schutze, 2001). Actually, Knowledge Discovery techniques are well integrated in many aspects of Language Technologies combining human background knowledge about the language with automatic approaches for modeling the 'soft' nature of ill structured data formulated in natural language. More on the usage of Language Technologies in knowledge management can be found in Cunningham and Bontcheva (2005).

It is also important to point out that scalability is one of the central issues in Knowledge Discovery, where one needs to be able to deal with real-life dataset volumes of the order of terabytes. Ontology construction is ultimately concerned with real-life data and on the Web today we talk about tens of billions of Web pages indexed by major search engines. Because of the exponential growth of data available in electronic form, especially on the Web, approaches where a large amount of human intervention is necessary, become inapplicable. Here we see a great potential for Knowledge Discovery with its focus on scalability.

The following subsections briefly describe some of the Knowledge Discovery techniques that can be used for addressing the ontology learning scenarios described in Section 2.5.

2.6.1. Unsupervised Learning

In the broader context, the Knowledge Discovery approach to ontology learning deals with some kind of data objects which need to have some kind of properties—these may be text documents, images, data records or some combination of them. From the perspective of using Knowledge Discovery methods for inducing concepts given the instances (ontology learning scenario 1 in Section 2.5), the important part is comparing ontological instances to each other. As document databases are the most common data type conceptualized in the form of ontologies, we can use methods developed in Information Retrieval and Text Mining research, for estimating similarity between documents as well as similarity between objects used within the documents (e.g., named entities, words, etc.)—these similarity measures can be used together with unsupervised learning algorithms, such as clustering algorithms, in an approach to forming an approximation of ontologies from document collections.

An approach to semi-automatic topic ontology construction from a collection of documents (ontology learning scenario 4 in Section 2.5) is proposed in Fortuna et al. (2005a). Ontology construction is seen as a process where the user is constructing the ontology and taking all the decisions while the computer provides suggestions for the topics (ontology concepts), and assists by automatically assigning documents to the topics, naming the topics, etc. The system is designed to take a set of documents and provide suggestions of possible ontology concepts (topics) and relations (sub-topic-of) based on the text of documents. The user can use the suggestions for concepts and their names, further split or refine the concepts, move a concept to another place in the ontology, explore instances of the concepts (in this case documents), etc. The system supports also extreme case where the user can ignore suggestions and manually construct the ontology. All this functionality is available through an interactive GUI-based environment providing ontology visualization and the ability to save the final ontology as RDF. There are two main methodological contributions introduced in this approach: (i) suggesting concepts as subsets of documents and (ii) suggesting naming of the concepts. Suggesting concepts based on the document collection is based on representing documents as word-vectors and applying *Document clustering or Latent Semantic Indexing (LSI)*. As ontology learning scenario 4 (described in Section 2.5) is one of the most important and demanding, in the remaining of this subsection we briefly describe both methods (clustering and LSI) for suggesting concepts. Turning to the second approach, naming of the concepts is based on proposing labels comprising the most common keywords (describing a subset of documents belonging to the topic), and alternatively on providing the most discriminative keywords (enabling classification of documents into the topic relative to the neighboring topics). Methods for document classification are briefly described in subsection 2.6.2.

Document clustering (Steinbach *et al.*, 2000) is based on a general data clustering algorithm adopted for textual data by representing each document as a word-vector, which for each word contains some weight proportional to the number of occurrences of the word (usually TFIDF weight as given in Equation (2.1)).

$$d^{(i)} = \text{TF}(W_i, d)\text{IDF}(W_i), \quad \text{where IDF}(W_i) = \log \frac{D}{\text{DF}(W_i)} \qquad (2.1)$$

where D is the number of documents; document frequency $\text{DF}(W)$ is the number of documents the word W occurred in at least once; and $\text{TF}(W, d)$ is the number of times word W occurred in document d. The exact formula used in different approaches may vary somewhat but the basic idea remains the same—namely, that the weighting is a measure of how frequently the given word occurs in the document at hand and of how common (or otherwise) the word is in an entire document collection.

The similarity of two documents is commonly measured by the cosine-similarity between the word-vector representations of the documents (see Equation (2.2)). The clustering algorithm group documents based on their similarity, putting similar documents in the same group. Cosine-similarity is commonly used also by some supervised learning algorithms for document categorization, which can be useful in populating topic ontologies (ontology learning scenario 3 in Section 2.5). Given a new document, cosine-similarity is used to find the most similar documents (e.g., using *k*-Nearest Neighbor algorithm (Mitchell, 1997)). Cosine-similarity between all the documents and the new document is used to find the k most similar documents whose categories (topics) are then used to assign categories to a new document. For documents d_i and d_j, the similarity is calculated as given in Equation (2.2). Note that the cosine-similarity between two identical documents is 1 and between two documents that share no words is 0.

$$\cos(d_i, d_j) = \frac{\sum_k d_{ik} d_{jk}}{\sqrt{\sum_l d_{il}^2 \sum_m d_{jm}^2}} \qquad (2.2)$$

Latent Semantic Indexing is a linear dimensionality reduction technique based on a technique from linear algebra called Singular Value

Decomposition. It uses a word-vector representation of text documents for extracting words with similar meanings (Deerwester *et al.*, 2001). It relies on the fact that two words related to the same topic more often cooccur together than words describing different topics. This can also be viewed as extraction of hidden semantic concepts or topics from text documents. The results of applying Latent Semantic Indexing on a document collection are fuzzy clusters of words each describing topics.

More precisely, in the process of extracting the hidden concepts first a term-document matrix A is constructed from a given set of text documents. This is a matrix having word-vectors of documents as columns. This matrix is decomposed using singular value decomposition so that $A \times USV^T$, where matrices U and V are orthogonal and S is a diagonal matrix with ordered singular values on the diagonal. Columns of the matrix U form an orthogonal basis of a subspace of the original space where vectors with higher singular values carry more information (by truncating singular values to only the k biggest values, we get the best approximation of matrix A with rank k). Because of this, vectors that form this basis can also be viewed as concepts or topics. Geometrically each basis vector splits the original space into two halves. By taking just the words with the highest positive or the highest negative weight in this basis vector, we get a set of words which best describe a concept generated by this vector. Note that each vector can generate two concepts; one is generated by positive weights and one by negative weights.

2.6.2. Semi-Supervised, Supervised, and Active Learning

Often it is too hard or too costly to integrate available background domain knowledge into fully automatic techniques. *Active Learning* and *Semi-supervised Learning* make use of small pieces of human knowledge for better guidance towards the desired model (e.g., an ontology). The effect is that we are able to reduce the amount of human effort by an order of magnitude while preserving the quality of results (Blum and Chawla, 2001). The main task of both the methods is to attach labels to unlabeled data (such as content categories to documents) by maximizing the quality of the label assignment and by minimizing the effort (human or computational).

A typical example scenario for using semi-supervised and active learning methods would be assigning content categories to uncategorized documents from a large document collection (e.g., from the Web or from a news source) as described in (Novak, 2004a). Typically, it is too costly to label each document manually—but there is some limited amount of human resource available. The task of active learning is to use the (limited) available user effort in the most efficient way, to assign high quality labels (e.g., in the form of content categories) to documents; semi-supervised learning, on the other hand, is applied when there are some initially labeled instances (e.g., documents with assigned topic categories) but no additional human resources are available. Finally, *supervised learning* is used when there is enough labeled data provided in advance and no additional human resources are available. All the three methods can be useful in populating ontologies (ontology learning scenario 3 in Section 2.5) using document categorization as well as in more sophisticated tasks such as inducing relations (ontology learning scenario 2 in Section 2.5), ontology generation and extension (ontology learning scenarios 4 and 5 in Section 2.5).

Supervised learning for text *document categorization* can be applied when a set of predefined topic categories, such as 'arts, education, science,' are provided as well as a set of documents labeled with those categories. The task is to classify new (previously unseen) documents by assigning each document one or more content categories (e.g., ontology concepts or relations). This is usually performed by representing documents as word-vectors and using documents that have already been assigned to the categories, to generate a model for assigning content categories to new documents (Jackson and Moulinier, 2002; Sebastiani, 2002). In the word-vector representation of a document, a vector of word frequencies is formed taking all the words occurring in all the documents (usually several thousands of words) and often applying some feature subset selection approach (Mladenic and Grobelnik, 2003). The representation of a particular document contains many zeros, as most of the words from the collection do not occur in a particular document. The categories can be organized into a topic ontology, for example, the MeSH ontology for medical subject headings or the Yahoo! hierarchy of Web documents that can be seen as a topic ontology.[1] Different Knowledge Discovery methods have been applied and evaluated on different document categorization problems. For instance, on the taxonomy of US patents, on Web documents organized in the Yahoo! Web directory (McCallum *et al.*, 1998; Mladenic, 1998; Mladenic and Grobelnik 2004), on the DMoz Web directory (Grobelnik and Mladenic 2005), on categorization of Reuters news articles (Koller and Sahami, 1997, Mladenic et al., 2004). Documents can also be related in ways other than common words (for instance, hyperlinks connecting Web documents) and these connections can be also used in document categorization (e.g., Craven and Slattery, 2001).

---

[1] The notion of a topic ontology is explored in detail in Chapter 7.

2.6.3. Stream Mining and Web Mining

Ontology updating is important not only because the ontology construction process is demanding and frequently requires further extension, but also because of the dynamic nature of the world (part of which is reflected in an ontology). The underlying data and the corresponding semantic structures change in time, the ontology gets used, etc. As a consequence, we would like to be able to adapt the ontologies accordingly. We refer to these kind of structures as 'dynamic ontologies' (ontology learning scenario 5 in Section 2.5). For most ontology updating scenarios, extensive human involvement in building models from the data is not economic, tending to be too costly, too inaccurate, and too slow.

A sub-field of Knowledge Discovery called *Stream Mining* addresses the issue of rapidly changing data. The idea is to be able to deal with the stream of incoming data quickly enough to be able to simultaneously update the corresponding models (e.g., ontologies), as the amount of data is too large to be stored: new evidence from the incoming data is incorporated into the model without storing the data. The underlying methods are based on the machine learning methods of *on-line learning*, where the model is built from the initially available data and updated regularly as more data becomes available.

Web-Mining, another sub-field of Knowledge Discovery, addresses Web data including three interleaved threads of research: Web *content* mining, Web *structure* mining, and Web *usage* mining. As ontologies are used in different applications and by different users, we can make an analogy between usage of ontologies and usage of Web pages. For instance, in Web usage mining (Chakrabarti, 2002), by analyzing frequencies of visits to particular Web pages and/or sequences of pages visited one after the other, one can consider restructuring the corresponding Web site or modeling the users behavior (e.g., in Internet shops, a certain sequence of visiting Web pages may be more likely to lead to a purchase than the other sequence). Using similar methods, we can analyze the usage patters of an ontology to identify parts of the ontology that are hardly used and reconsider their formulation, placement or existence. The appropriateness of Web usage mining methods for ontology updating still needs to be confirmed by further research.

2.6.4. Focused Crawling

An important step in ontology construction can be collecting the relevant data from the Web and using it for populating (ontology learning scenario 3 in Section 2.5) or updating the ontology (ontology learning scenario 5 in Section 2.5). Collecting data relevant for the existing ontology can also be used in some other phases of the semi-automatic ontology construction process, such as ontology evaluation or ontology refinement (phases 5 and 6, Section 2.4), for instance, via associating new instances to the existing ontology in a process called ontology grounding (Jakulin and Mladenic, 2005). In the case of topic ontologies (see Chapter 7), where the concepts correspond to topics and documents are linked to these topics through an appropriate relation such as hasSubject (Grobelnik and Mladenic 2005a), one can use the Web to collect documents on a predefined topic. In Knowledge Discovery, the approaches dealing with collecting documents based on the Web data are referred in the literature under the name *Focused Crawling* (Chakrabarti, 2002; Novak, 2004b). The main idea of these approaches is to use the initial 'seed' information given by the user to find similar documents by exploiting (1) background knowledge (ontologies, existing document taxonomies, etc.), (2) web topology (following hyperlinks from the relevant pages), and (3) document repositories (through search engines). The general assumption for most of the focused crawling methods is that pages with more closely related content are more inter-connected. In the cases where this assumption is not true (or we cannot reasonably assume it), we can still use the methods for selecting the documents through search engine querying (Ghani *et al.*, 2005). In general, we could say that focused crawling serves as a generic technique for collecting data to be used in the next stages of data processing, such as constructing (ontology learning scenario 4 in Section 2.5) and populating ontologies (ontology learning scenario 3 in Section 2.5).

2.6.5. Data Visualization

Visualization of data in general and also visualization of document collections is a method for obtaining early measures of data quality, content, and distribution (Fayyad *et al.*, 2001). For instance, by applying document visualization it is possible to get an overview of the content of a Web site or some other document collection. This can be useful especially for the first phases of semi-automatic ontology construction aiming at domain and data understanding (see Section 2.4). Visualization can be also used for visualizing an existing ontology or some parts thereof, which is potentially relevant for all the ontology learning scenarios defined in Section 2.5.

One general approach to document collection visualization is based on clustering of the documents (Grobelnik and Mladenic, 2002) by first representing the documents as word-vectors and performing k-means clustering on them (see Subsection 2.6.1). The obtained clusters are then represented as nodes in a graph, where each node in the graph is described by the set of most characteristic words in the

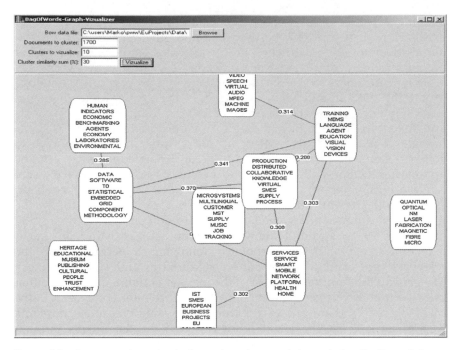

Figure 2.1   An example output of a system for graph-based visualization of document collection. The documents are 1700 descriptions of European research projects in information technology (5FP IST).

corresponding cluster. Similar nodes, as measured by their cosine-similarity (Equation (2.2)), are connected by a link. When such a graph is drawn, it provides a visual representation of the document set (see Figure 2.1 for an example output of the system). An alternative approach that provides different kinds of document corpus visualization is proposed in Fortuna *et al.*, 2005b). It is based on Latent Semantic Indexing, which is used to extract hidden semantic concepts from text documents and multidimensional scaling which is used to map the high dimensional space onto two dimensions. Document visualization can be also a part of more sophisticated tasks, such as generating a semantic graph of a document or supporting browsing through a news collection. For illustration, we provide two examples of document visualization that are based on Knowledge Discovery methods (see Figure 2.2 and Figure 2.3). Figure 2.2 shows an example of visualizing a single document via its semantic graph (Leskovec *et al.*, 2004). Figure 2.3 shows an example of visualizing news stories via visualizing relationships between the named entities that appear in the news stories (Grobelnik and Mladenic, 2004).

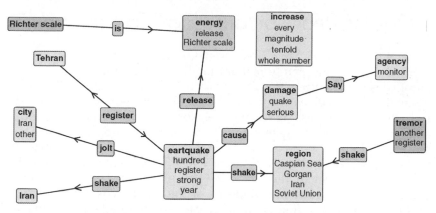

Figure 2.2 Visual representation of an automatically generated summary of a news story about earthquake. The summarization is based on deep parsing used for obtaining semantic graph of the document, followed by machine learning used for deciding which parts of the graph are to be included in the document summary.

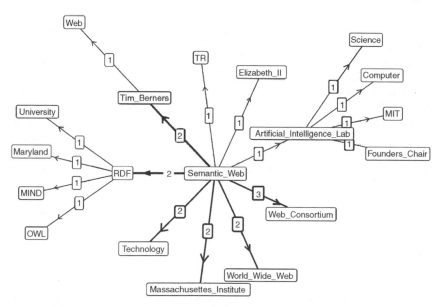

Figure 2.3 Visual representation of relationships (edges in the graph) between the named entities (vertices in the graph) appearing in a collection of news stories. Each edge shows intensity of comentioning of the two named entities. The graph is an example focused on the named entity 'Semantic Web' that was extracted from the 11.000 ACM Technology news stories from 2000 to 2004.

2.7. RELATED WORK ON ONTOLOGY CONSTRUCTION

Different approaches have been used for building ontologies, most of them to date using mainly manual methods. An approach to building ontologies was set up in the CYC project (Lenat and Guha, 1990), where the main step involved manual extraction of common sense knowledge from different sources. There have been some methodologies for building ontologies developed, again assuming a manual approach. For instance, the methodology proposed in (Uschold and King, 1995) involves the following stages: *identifying the purpose of the ontology* (why to build it, how will it be used, the range of the users), *building the ontology, evaluation and documentation*. Building of the ontology is further divided into three steps. The first is ontology capture, where key concepts and relationships are identified, a precise textual definition of them is written, terms to be used to refer to the concepts and relations are identified, the involved actors agree on the definitions and terms. The second step involves coding of the ontology to represent the defined conceptualization in some formal language (committing to some meta-ontology, choosing a representation language and coding). The third step involves possible integration with existing ontologies. An overview of methodologies for building ontologies is provided in Fernández (1999), where several methodologies, including the above described one, are presented and analyzed against the IEEE Standard for Developing Software Life Cycle Processes, thus viewing ontologies as parts of some software product. As there are some specifics to semi-automatic ontology construction compared to the manual approaches to ontology construction, the methodology that we have defined (see Section 2.4) has six phases. If we relate them to the stages in the methodology defined in Uschold and King (1995), we can see that the first two phases referring to domain and data understanding roughly correspond to *identifying the purpose of the ontology*, the next two phases (tasks definition and ontology learning) correspond to the stage of *building the ontology*, and the last two phases on ontology evaluation and refinement correspond to the *evaluation and documentation* stage.

Several workshops at the main Artificial Intelligence and Knowledge Discovery conferences (ECAI, IJCAI, KDD, ECML/PKDD) have been organized addressing the topic of ontology learning. Most of the work presented there addresses one of the following problems/tasks:

- Extending the existing ontology: Given an existing ontology with concepts and relations (commonly used is the English lexical ontology WordNet), the goal is to extend that ontology using some text, for example Web documents are used in (Agirre *et al.*, 2000). This can fit under the ontology learning scenario 5 in Section 2.5.

- Learning relations for an existing ontology: Given a collection of text documents and ontology with concepts, learn relations between the concepts. The approaches include learning taxonomic, for example isa, (Cimiano *et al.*, 2004) and nontaxonomic, for example 'hasPart' relations (Maedche and Staab, 2001) and extracting semantic relations from text based on collocations (Heyer *et al.*, 2001). This fits under the ontology learning scenario 2 in Section 2.5.
- Ontology construction based on clustering: Given a *collection of text documents*, split each document into sentences, parse the text and apply clustering for semi-automatic construction of an ontology (Bisson *et al.*, 2000; Reinberger and Spyns, 2004). Each cluster is labeled by the most characteristic words from its sentences or using some more sophisticated approach (Popescul and Ungar, 2000). Documents can be also used as a whole, without splitting them into sentences, and guiding the user through a semi-automatic process of ontology construction (Fortuna *et al.*, 2005a). The system provides suggestions for ontology concepts, automatically assigns documents to the concepts, proposed naming of the concepts, etc. In Hotho *et al.* (2003), the clustering is further refined by using WordNet to improve the results by mapping the found sentence clusters upon the concepts of a general ontology. The found concepts can be further used as semantic labels (XML tags) for annotating documents. This fits under the ontology learning scenario 4 in Section 2.5.
- Ontology construction based on semantic graphs: Given a *collection of text documents*, parse the documents; perform coreference resolution, anaphora resolution, extraction of subject-predicate-object triples, and construct semantic graphs. These are further used for learning summaries of the documents (Leskovec *et al.*, 2004). An example summary obtained using this approach is given in Figure 2.2. This can fit under the ontology learning scenario 4 in Section 2.5.
- Ontology construction from a collection of news stories based on named entities: Given a collection of news stories, represent it as a collection of graphs, where the nodes are named entities extracted from the text and relationships between them are based on the context and collocation of the named entities. These are further used for visualization of news stories in an interactive browsing environment (Grobelnik and Mladenic, 2004). An example output of the proposed approach is given in Figure 2.3. This can fit under the ontology learning scenario 4 in Section 2.5.

More information on ontology learning from text can be found in a collection of papers (Buitelaar *et al.*, 2005) addressing three perspectives: *methodologies* that have been proposed to automatically extract information from texts, *evaluation methods* defining procedures and metrics for a quantitative evaluation of the ontology learning task, and *application scenarios* that make ontology learning a challenging area in the context of real applications.

2.8. DISCUSSION AND CONCLUSION

We have presented several techniques from Knowledge Discovery that are useful for semi-automatic ontology construction. In that light, we propose to decompose the semi-automatic ontology construction process into several phases ranging from *domain and data understanding* through *task definition* via *ontology learning to ontology evaluation and refinement*. A large part of this chapter is dedicated to ontology learning. Several scenarios are identified in the ontology learning phase depending on different assumptions regarding the provided input data and the expected output: inducing concepts, inducing relations, ontology population, ontology construction, and ontology updating/extension. Different groups of Knowledge Discovery techniques are briefly described including unsupervised learning, semi-supervised, supervised and active learning, on-line learning and web-mining, focused crawling, data visualization. In addition to providing brief description of these techniques, we also relate them to different ontology learning scenarios that we identified.

Some of the described Knowledge Discovery techniques have already been applied in the context of semi-automatic ontology construction, while others still need to be adapted and tested in that context. A challenge for future research is setting up evaluation frameworks for assessing contribution of these techniques to specific tasks and phases of the ontology construction process. In that light, we briefly describe some existing approaches to ontology construction and point to the original papers that provide more information on the approaches, usually including some evaluation of their contribution and performance on the specific tasks. We also related existing work on learning ontologies to different ontology learning scenarios that we have identified. Our hope is that this chapter in addition to contributing by proposing a methodology for semi-automatic ontology construction and description of some relevant Knowledge Discovery techniques also shows potential for future research and triggers some new ideas related to the usage of Knowledge Discovery techniques for ontology construction.

ACKNOWLEDGMENTS

This work was supported by the Slovenian Research Agency and the IST Programme of the European Community under SEKT Semantically Enabled Knowledge Technologies (IST-1-506826-IP) and PASCAL Network of Excellence (IST-2002-506778). This publication only reflects the authors' views.

REFERENCES

Agirre E, Ansa O, Hovy E, Martínez D. 2000. Enriching very large ontologies using the WWW. In Proceedings of the First Workshop on Ontology Learning OL-2000. The 14th European Conference on Artificial Intelligence ECAI-2000.

Bisson G, Nédellec C, Cañamero D. 2000. Designing clustering methods for ontology building: The Mo'K workbench. In Proceedings of the First Workshop on Ontology Learning OL-2000. The 14th European Conference on Artificial Intelligence ECAI-2000.

Bloehdorn S, Haase P, Sure Y, Voelker J, Bevk M, Bontcheva K, Roberts I. 2005. Report on the integration of ML, HLT and OM. SEKT Deliverable D.6.6.1, July 2005.

Blum A, Chawla S. 2001. Learning from labelled and unlabelled data using graph mincuts. Proceedings of the 18th International Conference on Machine Learning, pp 19–26.

Buitelaar P, Cimiano P, Magnini B. 2005. Ontology learning from text: Methods, applications and evaluation. frontiers in Artificial Intelligence and Applications, IOS Press.

Brank J, Grobelnik M, Mladenic D. 2005. A survey of ontology evaluation techniques. Proceedings of the 8th International multi-conference Information Society IS-2005, Ljubljana: Institut "Jožef Stefan", 2005.

Chakrabarti S. 2002. *Mining the Web: Analysis of Hypertext and Semi Structured Data*. Morgan Kaufmann.

Chapman P, Clinton J, Kerber R, Khabaza T, Reinartz T, Shearer C, Wirth R. 2000. CRISP-DM 1.0: Step-by-step data mining guide.

Cimiano P, Pivk A, Schmidt-Thieme L, Staab S. 2004. Learning taxonomic relations from heterogeneous evidence. In Proceedings of ECAI 2004 Workshop on Ontology Learning and Population.

Craven M, Slattery S. 2001. Relational learning with statistical predicate invention: better models for hypertext. *Machine Learning* 43(1/2):97–119.

Cunningham H, Bontcheva K. 2005. Knowledge management and human language: crossing the chasm. *Journal of Knowledge Management*.

Deerwester, S., Dumais, S., Furnas, G., Landuer, T., Harshman, R., (2001). Indexing by Latent Semantic Analysis.

Duda RO, Hart PE, Stork DG 2000. *Pattern Classification* (2nd edn). John Wiley & Sons, Ltd.

Ehrig M, Haase P, Hefke M, Stojanovic N. 2005. Similarity for ontologies—A comprehensive framework. Proceedings of 13th European Conference on Information Systems, May 2005.

Fayyad, U., Grinstein, G. G. and Wierse, A. (eds.), (2001). Information Visualization in Data Mining and Knowledge Discovery, Morgan Kaufmann.

Fayyad U, Piatetski-Shapiro G, Smith P, Uthurusamy R (eds). 1996. *Advances in Knowledge Discovery and Data Mining*. MIT Press: Cambridge, MA, 1996.

Fernández LM. 1999. Overview of methodologies for building ontologies. In Proceedings of the IJCAI-99 workshop on Ontologies and Problem-Solving Methods (KRR5).

Fortuna B, Mladenic D, Grobelnik M. 2005a. Semi-automatic construction of topic ontology. Proceedings of the ECML/PKDD Workshop on Knowledge Discovery for Ontologies.

Fortuna B, Mladenic D, Grobelnik M. 2005b. Visualization of text document corpus. *Informatica journal* 29(4):497–502.

Ghani R, Jones R, Mladenic D. 2005. Building minority language corpora by learning to generate web search queries. *Knowledge and information systems* 7:56–83.

Grobelnik M, Mladenic D. 2002. Efficient visualization of large text corpora. Proceedings of the seventh TELRI seminar. Dubrovnik, Croatia.

Grobelnik M, Mladenic D. 2004. Visualization of news articles. *Informatica Journal* 28:(4).

Grobelnik M, Mladenic D. 2005. Simple classification into large topic ontology of Web documents. *Journal of Computing and Information Technology—CIT 13* 4:279–285.

Grobelnik M, Mladenic D. 2005a. Automated knowledge discovery in advanced knowledge management. *Journal of Knowledge Management.*

Hand DJ, Mannila H, Smyth P. 2001. *Principles of Data Mining (Adaptive Computation and Machine Learning).* MIT Press.

Hastie T, Tibshirani R, Friedman JH. 2001. *The Elements of Statistical Learning*: *Data Mining, Inference, and Prediction.* Springer Series in Statistics. Springer Verlag.

Heyer G, Läuter M, Quasthoff U, Wittig T, Wolff C. 2001. Learning Relations using Collocations. In Proceedings of IJCAI-2001 Workshop on Ontology Learning.

Hotho A, Staab S, Stumme G. 2003. Explaining text clustering results using semantic structures. In Proceedings of ECML/PKDD 2003, LNAI 2838, Springer Verlag, pp 217–228.

Jackson P, Moulinier I. 2002. *Natural Language Processing for Online Applications*: *Text Retrieval, Extraction, and Categorization.* John Benjamins Publishing Co.

Jakulin A, Mladenic D. 2005. Ontology grounding. Proceedings of the 8th International multi-conference Information Society IS-2005, Ljubljana: Institut "Jožef Stefan", 2005.

Koller D, Sahami M. 1997. Hierarchically classifying documents using very few words. Proceedings of the 14th International Conference on Machine Learning ICML-97, Morgan Kaufmann, San Francisco, CA, pp 170–178.

Leskovec J, Grobelnik M, Milic-Frayling N. 2004. Learning sub-structures of document semantic graphs for document summarization. In Workshop on Link Analysis and Group Detection (LinkKDD2004). The Tenth ACM SIGKDD International Conference on Knowledge Discovery and Data Mining.

Maedche A, Staab S. 2001. Discovering conceptual relations from text. In Proceedings of ECAI'2000, pp 321–325.

Manning CD, Schutze H. 2001. *Foundations of Statistical Natural Language Processing.* The MIT Press: Cambridge, MA.

McCallum A, Rosenfeld R, Mitchell T, Ng A. 1998. Improving text classification by shrinkage in a hierarchy of classes. Proceedings of the 15th International Conference on Machine Learning ICML-98, Morgan Kaufmann, San Francisco, CA.

Mitchell TM. 1997. *Machine Learning.* The McGraw-Hill Companies, Inc.

Mladenic D. 1998. Turning Yahoo into an Automatic Web-Page Classifier. Proceedings of 13th European Conference on Artificial Intelligence (ECAI'98, John Wiley & Sons, Ltd), pp 473–474.

Mladenic D, Brank J, Grobelnik M, Milic-Frayling N. 2002. Feature selection using linear classifier weights: Interaction with classification models, SIGIR-2002.

Mladenic D, Grobelnik M. 2003. Feature selection on hierarchy of web documents. *Journal of Decision support systems* 35:45–87.

Mladenic D, Grobelnik M. 2004. Mapping documents onto web page ontology. In *Web Mining: From Web to Semantic Web,* (Berendt B, Hotho A, Mladenic D, Someren MWV, Spiliopoulou M, Stumme G (eds). Lecture notes in artificial inteligence, Lecture notes in computer science, Vol. 3209. Springer: Berlin; Heidelberg; New York, 2004; 77–96.

Novak B. 2004a. Use of unlabeled data in supervised machine learning. Proceedings of the 7th International multi-conference Information Society IS-2004, Ljubljana: Institut "Jožef Stefan", 2004.

Novak B. 2004b. A survey of focused web crawling algorithms. Proceedings of the 7th International multi-conference Information Society IS-2004, Ljubljana: Institut "Jožef Stefan", 2004.

Popescul A, Ungar LH. 2000. Automatic labeling of document clusters. Department of Computer and Information Science, University of Pennsylvania, unpublished paper available from http://www.cis.upenn.edu/~popescul/Publications/popescul00labeling.pdf

Reinberger M-L, Spyns P. 2004. Discovering Knowledge in Texts for the learning of DOGMA-inspired ontologies. In Proceedings of ECAI 2004 Workshop on Ontology Learning and Population.

Sebastiani F. 2002. Machine learning for automated text categorization. *ACM Computing Surveys*.

Steinbach M, Karypis G, Kumar V. 2000. A comparison of document clustering techniques. Proceedings of KDD Workshop on Text Mining (Grobelnik M, Mladenić D, Milic-Frayling N (eds)), Boston, MA, USA, pp 109–110.

Uschold M, King M. 1995. Towards a methodology for building ontologies. In Workshop on Basic Ontological Issues in Knowledge Sharing. International Joint Conference on Artificial Intelligence, 1995. Also available as AIAI-TR-183 from AIAI, the University of Edinburgh.

van Rijsbergen CJ. 1979. *Information Retrieval* (2nd edn). Butterworths, London.

Witten IH, Frank E. 1999. *Data Mining: Practical Machine Learning Tools and Techniques with Java Implementations*. Morgan Kaufmann.

3

Semantic Annotation and Human Language Technology

Kalina Bontcheva, Hamish Cunningham, Atanas Kiryakov and
Valentin Tablan

3.1. INTRODUCTION

Gartner reported in 2002 that for at least the next decade more than 95%
of human-to-computer information input will involve textual language.
They also report that by 2012, taxonomic and hierarchical knowledge
mapping and indexing will be prevalent in almost all information-rich
applications. There is a tension here: between the increasingly rich
semantic models in IT systems on the one hand, and the continuing
prevalence of human language materials on the other. The process of
tying semantic models and natural language together is referred to as
Semantic Annotation. This process may be characterised as the dynamic
creation of inter-relationships between ontologies (shared conceptualisa-
tions of domains) and documents of all shapes and sizes in a
bidirectional manner covering creation, evolution, population and documentation of ontological models. Work in the Semantic Web (Berners-
Lee, 1999; Davies *et al.*, 2002; Fensel *et al.*, 2002) (see also other chapters in
this volume) has supplied a standardised, web-based suite of languages
(e.g., Dean *et al.*, 2004) and tools for the representation of ontologies and
the performance of inferences over them. It is probable that these
facilities will become an important part of next-generation IT applications, representing a step up from the taxonomic modelling that is now
used in much leading-edge IT software. Information Extraction (IE), a

Semantic Web Technologies: Trends and Research in Ontology-based Systems
John Davies, Rudi Studer, Paul Warren © 2006 John Wiley & Sons, Ltd

form of natural language analysis, is becoming a central technology to link Semantic Web models with documents as part of the process of Metadata Extraction.

The Semantic Web aims to add a machine tractable, repurposeable layer to complement the existing web of natural language hypertext. In order to realise this vision, the creation of semantic annotation, the linking of web pages to ontologies and the creation, evolution and interrelation of ontologies must become automatic or semi-automatic processes.

In the context of new work on distributed computation, Semantic Web Services (SWSs) go beyond current services by adding ontologies and formal knowledge to support description, discovery, negotiation, mediation and composition. This formal knowledge is often strongly related to informal materials. For example, a service for multimedia content delivery over broadband networks might incorporate conceptual indices of the content, so that a smart VCR (such as next generation TiVO) can reason about programmes to suggest to its owner. Alternatively, a service for B2B catalogue publication has to translate between existing semi-structured catalogues and the more formal catalogues required for SWS purposes. To make these types of services cost-effective, we need automatic knowledge harvesting from all forms of content that contain natural language text or spoken data.

Other services do not have this close connection with informal content, or will be created from scratch using Semantic Web authoring tools. For example, printing or compute cycle or storage services. In these cases the opposite need is present: to document services for the human reader using natural language generation.

An important aspect of the world wide web revolution is that it has been based largely on human language materials, and in making the shift to the next generation knowledge-based web, human language will remain key. Human Language Technology (HLT) involves the analysis, mining and production of natural language. HLT has matured over the last decade to a point at which robust and scaleable applications are possible in a variety of areas, and new projects like SEKT in the Semantic Web area are now poised to exploit this development.

Figure 3.1 illustrates the way in which Human Language Technology can be used to bring together the natural language upon which the current web is mainly based and the formal knowledge at the basis of the Semantic Web. Ontology-Based IE and Controlled Language IE are discussed in this chapter, whereas Natural Language Generation is covered in Chapter 8 on Knowledge Access.

The chapter is structured as follows. Section 3.2 provides an overview of Information Extraction (IE) and the problems it addresses. Section 3.3 introduces the problem of semantic annotation and shows why it is harder than the issues addressed by IE. Section 3.4 surveys some applications of IE to semantic annotation and discusses the problems

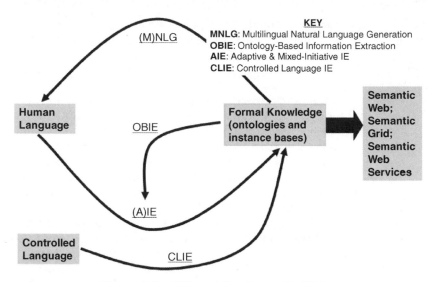

Figure 3.1   HLT and the Semantic Web.

faced, thus justifying the need for the so-called Ontology-Based IE approaches. Section 3.5 presents a number of these approaches, including some graphical user interfaces. Controlled Language IE (CLIE) is then presented as a high-precision alternative to information extraction from unrestricted, ambiguous text. The chapter concludes with a discussion and outlines future work.

3.2. INFORMATION EXTRACTION: A BRIEF INTRODUCTION

Information extraction (IE) is a technology based on analysing natural language in order to extract snippets of information. The process takes texts (and sometimes speech) as input and produces fixed format, unambiguous data as output. This data may be used directly for display to users, or may be stored in a database or spreadsheet for later analysis, or may be used for indexing purposes in information retrieval (IR) applications such as internet search engines like Google.

IE is quite different from IR:

- an IR system finds relevant texts and presents them to the user;
- an IE application analyses texts and present only the specific information from them that the user is interested in.

For example, a user of an IR system wanting information on trade group formations in agricultural commodities markets would enter a list of relevant words and receive in return a set of documents (e.g., newspaper articles) which contain likely matches. The user would then read the documents and extract the requisite information themselves. They might then enter the information in a spreadsheet and produce a chart for a report or presentation. In contrast, an IE system would automatically populate the spreadsheet directly with the names of relevant companies and their groupings.

There are advantages and disadvantages with IE in comparison to IR. IE systems are more difficult and knowledge-intensive to build, and are to varying degrees tied to particular domains and scenarios. IE is more computationally intensive than IR. However, in applications where there are large text volumes IE is potentially much more efficient than IR because of the possibility of dramatically reducing the amount of time people spend reading texts. Also, where results need to be presented in several languages, the fixed-format, unambiguous nature of IE results makes this relatively straightforward in comparison with providing the full translation facilities needed for interpretation of multilingual texts found by IR.

Useful overview sources for further details on IE include: Cowie and Lehnert (1996), Appelt (1999), Cunningham (2005), Gaizauskas and Wilks (1998) and Pazienza (2003).

3.2.1. Five Types of IE

IE is about finding five different types of information in natural language text:

1. Entities: things in the text, for example people, places, organisations, amounts of money, dates, etc.
2. Mentions: all the places that particular entities are referred to in the text.
3. Descriptions of the entities present.
4. Relations between entities.
5. Events involving the entities.

For example, consider the text:

> 'Ryanair announced yesterday that it will make Shannon its next European base, expanding its route network to 14 in an investment worth around €180m. The airline says it will deliver 1.3 million passengers in the first year of the agreement, rising to two million by the fifth year'.

To begin with, IE will discover that 'Shannon' and 'Ryanair' are entities (of types location and company, perhaps), then, via a process of reference resolution, will discover that 'it' and 'its' in the first sentence refer to Ryanair (or are mentions of that company), and 'the airline' and 'it' in the second sentence also refer to Ryanair. Having discovered the mentions descriptive information can be extracted, for example that Shannon is a European base. Finally relations, for example that Shannon will be a base of Ryanair, and events, for example that Ryanair will invest €180 million in Shannon.

These various types of IE provide progressively higher-level information about texts. They are described in more detail below; for a thorough discussion and examples see Cunningham (2005).

3.2.2. Entities

The simplest and most reliable IE technolog is entity recognition, which we will abbreviate NE following the original Message Understanding Conference (MUC) definitions (SAIC, 1998) NE systems identify all the names of people, places, organisations, dates, amounts of money, etc.

All things being equal, NE recognition can be performed at up to around 95 % accuracy. Given that human annotators do not perform to the 100 % level (measured by inter-annotator comparisons), NE recognition can now be said to function at human performance levels, and applications of the technology are increasing rapidly as a result.

The process is weakly domain-dependent, that is changing the subject matter of the texts being processed from financial news to other types of news would involve some changes to the system, and changing from news to scientific papers would involve quite large changes.

3.2.3. Mentions

Finding the mentions of entities involves using of coreference resolution (CO) to identify identity relations between entities in texts. These entities are both those identified by NE recognition and anaphoric references to those entities. For example, in:

'Alas, poor Yorick, I knew him Horatio'.

coreference resolution would tie 'Yorick' with 'him' (and 'I' with Hamlet, if sufficient information was present in the surrounding text).

This process is less relevant to end users than other IE tasks (i.e. whereas the other tasks produce output that is of obvious utility for the application user, this task is more relevant to the needs of the application developer). For text browsing purposes, we might use CO to highlight all occurrences of the same object or provide hypertext links between them. CO technology might also be used to make links between documents. The main significance of this task, however, is as a building block for TE and ST (see below). CO enables the association of descriptive information scattered across texts with the entities to which it refers.

CO breaks down into two sub-problems: anaphoric resolution (e.g., 'I' with Hamlet); proper-noun resolution. Proper-noun coreference identification finds occurences of same object represented with different spelling or compounding, for example 'IBM', 'IBM Europe', 'International Business Machines Ltd', $\cdots$). CO resolution is an imprecise process, particularly when applied to the solution of anaphoric reference. CO results vary widely; depending on domain perhaps only 50–60 % may be relied upon. CO systems are domain dependent.

3.2.4. Descriptions

The description extraction task builds on NE recognition and coreference resolution, associating descriptive information with the entities. To match the original MUC definitions as before, we will abbreviate this task as 'TE'. For example, in a news article the 'Bush administration' can be also referred to as 'government officials'—the TE task discovers this automatically and adds it as an alias.

Good scores for TE systems are around 80 % (on similar tasks humans can achieve results in the mid 90s, so there is some way to go). As in NE recognition, the production of TEs is weakly domain dependent, that is changing the subject matter of the texts being processed from financial news to other types of news would involve some changes to the system, and changing from news to scientific papers would involve quite large changes.

3.2.5. Relations

As described in Appelt (1999), 'The template relation task requires the identification of a small number of possible relations between the template elements identified in the template element task. This might be, for example, an employee relationship between a person and a company, a family relationship between two persons, or a subsidiary relationship between two companies. Extraction of relations among entities is a central feature of almost any information extraction task, although the possibilities in real-world extraction tasks are endless'. In general good template relation (TR) system scores reach around 75 %. TR is a weakly domain dependent task.

3.2.6. Events

Finally, event extraction, which is abbreviated ST, for scenario template, the MUC style of representing information relating to events. (In some ways STs are the prototypical outputs of IE systems, being the original task for which the term was coined.) They tie together TE entities and TR relations into event descriptions. For example, TE may have identified Mr Smith and Mr Jones as person entities and a company present in a news article. TR would identify that these people work for the company. ST then identifies facts such as that they signed a contract on behalf of the company with another supplier company.

ST is a difficult IE task; the best MUC systems score around 60 %. The human score can be as low as around 80+ %, which illustrates the complexity involved. These figures should be taken into account when considering appropriate applications of ST technology. Note that it is possible to increase precision at the expense of recall: we can develop ST systems that do not make many mistakes, but that miss quite a lot of occurrences of relevant scenarios. Alternatively we can push up recall and miss less, but at the expense of making more mistakes.

The ST task is both domain dependent and, by definition, tied to the scenarios of interest to the users. Note however that the results of NE, TR and TE feed into ST, thus leading to an overall lower score due to a certain compounding of errors from the earlier stages.

3.3. SEMANTIC ANNOTATION

Semantic annotation is a specific metadata generation and usage schema aiming to enable new information access methods and to enhance existing ones. The annotation scheme offered here is based on the understanding that the information discovered in the documents by an IE system constitute an important part of their semantics. Moreover, by using text redundancy and external or background knowledge, this information can be connected to formal descriptions, that is, ontologies, and thus provide semantics and connectivity to the web.

The task of realising the vision of the Semantic Web will be much helped, if the following basic tasks can be properly defined and solved:

1. Formally annotate and hyperlink (references to) entities and relations in textual (parts of) documents.
2. Index and retrieve documents with respect to entities/relations referred to.

The first task could be seen as a combination of a basic press-clipping exercise, a typical IE task, and automatic hyper-linking. The resulting annotations represent a method for document enrichment and presenta-tion, the results of which can be further used to enable other access methods (see Chapter 8 on Knowledge Access). The second task is just a modification of the classical IR task—documents are retrieved on the basis of relevance to entities or relations instead of words. However the basic assumption is quite similar—a document is characterised by the bag of tokens constituting its content, disregarding its structure. While the basic IR approach considers word stems as tokens, there has been considerable effort in the last decade towards using word-senses or lexical concepts (see Mahesh *et al.*, 1999; Voorhees *et al.*, 1998) for

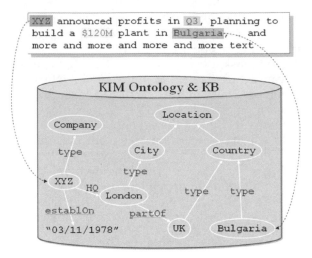

Figure 3.2 Semantic annotation.

indexing and retrieval. Similarly, entities and relations can be seen as a special sort of a token to be indexed and retrieved.

In a nutshell, Semantic Annotation is about assigning to entities and relations in the text links to their semantic descriptions in an ontology (as shown in Figure 3.2). This sort of semantic metadata provides both class and instance information about the entities/relations.

Most importantly, automatic semantic annotation enables many new applications: highlighting, semantic search, categorisation, generation of more advanced metadata, smooth traversal between unstructured text and formal knowledge. Semantic annotation is applicable to any kind of content—web pages, regular (nonweb) documents, text fields in databases, video, audio, etc.

3.3.1. What is Ontology-Based Information Extraction

Ontology-Based IE (OBIE) is the technology used for semantic annotation. One of the important differences between traditional IE and OBIE is the use of a formal ontology as one of the system's resources. OBIE may also involve reasoning.

Another substantial difference of the semantic IE process from the traditional one is the fact that it not only finds the (most specific) type of the extracted entity, but it also identifies it, by linking it to its semantic description in the instance base. This allows entities to be traced across documents and their descriptions to be enriched through the IE process. When compared to the 'traditional' IE tasks discussed in Section 3.2, the first stage corresponds to the NE task and the second stage corresponds to the CO (coreference) task. Given the lower performance achievable on the CO task, semantic IE is in general a much harder task.

OBIE poses two main challenges:

- the identification of instances from the ontology in the text;
- the automatic population of ontologies with new instances in the text.

3.3.1.1. Identification of Instances From the Ontology

If an ontology is already populated with instances, the task of an OBIE system may be simply to identify instances from the ontology in the text. Similar methodologies can be used for this as for traditional IE systems, using an ontology rather than a flat gazetteer. For rule-based systems, this is relatively straightforward. For learning-based systems, however, this is more problematic because training data is required. Collecting such training data is, however, likely to be a large bottleneck. Unlike traditional IE systems for which training data exists in domains like news texts in plentiful form, thanks to efforts from MUC, ACE (ACE, 2004) and other collaborative and/or competitive programs, there is a dearth of material currently available for semantic web applications. New training data needs to be created manually or semi-automatically, which is a time-consuming and onerous task, although systems to aid such metadata creation are currently being developed.

3.3.1.2. Automatic Ontology Population

In this task, an OBIE application identifies instances in the text belonging to concepts in a given ontology, and adds these instances to the ontology in the correct location. It is important to note that instances may appear in more than one location in the ontology because of the multidimensional nature of many ontologies and/or ambiguity which cannot or should not be resolved at this level (see e.g., Felber, 1984; Bowker, 1995 for a discussion).

3.4. APPLYING 'TRADITIONAL' IE IN SEMANTIC WEB APPLICATIONS

In this section, we give a brief overview of some current state-of-the-art systems which apply traditional IE techniques for semantic web applications such as annotating web pages with metadata. Unlike ontology-based IE applications, these do not incorporate ontologies into the system, but either use ontologies as a bridge between the IE system and the final annotation (as with AERODAML) or rely on the user to provide the relevant information through manual annotation (as with the Amilcare-based tools).

3.4.1. AeroDAML

AeroDAML (Kogut and Holmes, 2001) is an annotation tool created by Lockheed Martin, which applies IE techniques to automatically generate DAML annotations from web pages. The aim is to provide naive users with a simple tool to create basic annotations without having to learn about ontologies, in order to reduce time and effort and to encourage people to semantically annotate their documents. AeroDAML links most proper nouns and common types of relations with classes and properties in a DAML ontology.

There are two versions of the tool: a web-enabled version which uses a default generic ontology, and a client-server version which supports customised ontologies. In both cases, the user enters a URI (for the former) and a filename (for the latter) and the system returns the DAML annotation for the webpage or document. It provides a drag-and-drop tool to create static (manual) ontology mappings, and also includes some mappings to predefined ontologies.

AeroDAML consists of the AeroText IE system, together with components for DAML generation. A default ontology which directly correlates to the linguistic knowledge base used by the extraction process is used to translate the extraction results into a corresponding RDF model that uses the DAML+OIL syntax. This RDF model is then serialised to produce the final DAML annotation. The AeroDAML ontology comprises two layers: a base layer comprising the common knowledge base of AeroText, and an upper layer based on WordNet. AeroDAML can generate annotations consisting of instances of classes such as common nouns and proper nouns, and properties, of types such as coreference, Organisation to Location, Person to Organization.

3.4.2. Amilcare

Amilcare (Ciravegna and Wilks, 2003) is an IE system which has been integrated in several different annotation tools for the Semantic Web. It uses machine learning (ML) to learn to adapt to new domains and applications using only a set of annotated texts (training data). It has been adapted for use in the Semantic Web by simply monitoring the kinds of annotations produced by the user in training, and learning how to reproduce them. The traditional version of Amilcare adds XML annotations to documents (inline markup); the Semantic Web version leaves the original text unchanged and produces the extracted information as triples of the form ⟨annotation, startPosition, endPosition⟩ (stand-off markup). This means that it is left to the annotation tool and not the IE system to decide on the format of the ultimate annotations produced.

In the Semantic Web version, no knowledge of IE is necessary; the user must simply define a set of annotations, which may be organised as an ontology where annotations are associated with concepts and relations. The user then manually annotates the text using some interface connected to Amilcare, as described in the following systems. Amilcare works by preprocessing the texts using GATE's IE system ANNIE (Cunningham *et al.*, 2002), and then uses a supervised machine learning algorithm to induce rules from the training data.

3.4.3. MnM

MnM (Motta *et al.*, 2002) is a semantic annotation tool which provides support for annotating web pages with semantic metadata. This support is semi-automatic, in that the user must provide some initial training information by manually annotating documents before the IE system (Amilcare) can take over. It integrates a web browser, an ontology editor, and tools for IE, and has been described as 'an early example of next-generation ontology editors' (Motta *et al.*, 2002), because it is web-based and provides facilities for large-scale semantic annotation of web pages. It aims to provide a simple system to perform knowledge extraction tasks at a semi-automatic level.

There are five main steps to the procedure:

- the user browses the web;
- the user manually annotates his chosen web pages;
- the system learns annotation rules;
- the system tests the rules learnt;
- the system takes over automatic annotation, and populate ontologies with the instances found. The ontology population process is semi-automatic and may require intervention from the user.

3.4.4. S-Cream

S-CREAM (Semi-automatic CREAtion of Metadata) (Handschuh *et al.*, 2002) is a tool which provides a mechanism for automatically annotating texts, given a set of training data which must be manually created by the user. It uses a combination of two tools: Onto-O-Mat, a manual annotation tool which implements the CREAM framework for creating relational metadata (Handschuh *et al.*, 2001), and Amilcare.

As with MnM, S-CREAM is trainable for different domains, provided that the user creates the necessary training data. It essentially works by aligning conceptual markup (which defines relational metadata) provided by Ont-O-Mat with semantic markup provided by Amilcare. This problem is not trivial because the two representations may be very different. Relational metadata may provide information about relationships between instances of classes, for example that a certain hotel is located in a certain city. S-CREAM thus supports metadata creation with the help of a traditional IE system, and also provides other functionalities such as web crawler, document management system, and a meta-ontology.

3.4.5. Discussion

One of the problems with these annotation tools is that they do not provide the user with a way to customise the integrated language technology directly. While many users would not need or want such customisation facilities, users who already have ontologies with rich instance data will benefit if they can make this data available to the IE components. However, this is not possible when 'traditional' IE methods like Amilcare are used because they are not aware of the existence of the user's ontology.

The more serious problem however, as discussed in the S-CREAM system (Handschuh *et al.*, 2002), is that there is often a gap between the annotations and their types produced by IE and the classes and properties in the user's ontology. The proposed solution is to write some kind of rules, such as logical rules, to achieve this. For example, an IE system would typically annotate London and UK as locations, but extra rules are needed to specify that there is a containment relationship between the two (for other examples see (Handschuh *et al.*, 2002)). However, rule writing of the proposed kind is too difficult for most users and a new solution is needed to bridge this gap.

Ontology-Based IE systems for semantic annotation, to be discussed next, address both problems:

- The ontology is used as a resource during the IE process and therefore it can benefit from existing data such as names of customers from a billing database.
- Instance disambiguation is performed as part of the semantic annotation process, thus removing the need for user-written rules.

3.5. ONTOLOGY-BASED IE

3.5.1. Magpie

Magpie (Domingue *et al.*, 2004) is a suite of tools which supports the interpretation of webpages and 'collaborative sense-making'. It annotates webpages with metadata in a fully automatic fashion and needs no manual intervention by matching the text against instances in the ontology. It automatically populates an ontology from relevant web sources, and can be used with different ontologies. The principle behind it is that it uses an ontology to provide a very specific and personalised viewpoint of the webpages the user wishes to browse. This is important because different users often have different degrees of knowledge and/ or familiarity with the information presented, and have different browsing needs and objectives.

Magpie's main limitation is that it does not perform automatic population of the ontology with new instances, that is, it is restricted only to matching mentions of already existing instances.

3.5.2. Pankow

The PANKOW system (Pattern-based Annotation through Knowledge on the Web) (Cimiano *et al.*, 2004) exploits surface patterns and the redundancy on the Web to categorise automatically instances from text with respect to a given ontology. The patterns are phrases like: the ⟨INSTANCE⟩ ⟨CONCEPT⟩ (e.g. the Ritz hotel) and ⟨INSTANCE⟩ is a ⟨CONCEPT⟩ (e.g., Novotel is a hotel). The system constructs patterns by identifying all proper names in the text (using a part-of-speech tagger) and combining each one of them with each of the 58 concepts from their tourism ontology into a hypothesis. Each hypothesis is then checked against the Web via Google queries and the number of hits is used as a measure of the likelihood of this pattern being correct.

The system's best performance on this task in fully automatic mode is 24.9 % while the human performance is 62.09 %. However, when the system is used in semi-automatic mode, that is, it suggests the top five most likely concepts and the user chooses among them, then the performance goes up to 49.56 %.

The advantages of this approach are that it does not require any text processing (apart from POS tagging) or any training data. All the information comes from the web. However, this is also a major disadvantage because the method does not compare the context in which the proper name occurs in the document to the contexts in which it occurs on the Web, thus making it hard to classify instances with the same name that belong to different classes in different contexts (e.g., Niger can be a river, state, country, etc.). On the other hand, while IE systems are more costly to set up, they can take context into account when classifying proper names.

3.5.3. SemTag

The SemTag system (Dill *et al.*, 2003) performs large-scale semantic annotation with respect to the TAP ontology.[1] It first performs a lookup

---

[1]http://tap.stanford.edu/tap/papers.thml

phase annotating all possible mentions of instances from the TAP ontology. In the second, disambiguation phase, SemTag uses a vector-space model to assign the correct ontological class or determine that this mention does not correspond to a class in TAP. The disambiguation is carried out by comparing the context of the current mention with the contexts of instances in TAP with compatible aliases, using a window of 10 words either side of the mention.

The TAP ontology, which contains about 65,000 instances, is very similar in size and structure to the KIM Ontology and instance base discussed in Section 5.5. (e.g., each instance has a number of lexical aliases). One important characteristic of both ontologies is that they are very lightweight and encode only essential properties of concepts and instances. In other words, the goal is to cover frequent, commonly known and searched for instances (e.g., capital cities, names of pre-sidents), rather than to encode an extensive set of axioms enabling deep, Cyc-style reasoning. As reported by (Mahesh *et al.*, 1996), the heavy-weight logical approach undertaken in Cyc is not appropriate for many NLP tasks.

The SemTag system is based on a high-performance parallel architec-ture -Seeker, where each node annotates about 200 documents per second. The demand for such parallelism comes from the big volumes of data that need to be dealt with in many applications and make automatic semantic annotation the only feasible option. A parallel architecture of a similar kind is currently under development for KIM and, in general, it is an important ingredient of large-scale automatic annotation approaches.

3.5.4. Kim

The Knowledge and Information Management system (KIM) is a product of OntoText Lab which is currently being used and further developed in SEKT (Kiryakov *et al.*, 2005).

KIM is an extensible platform for semantics-based knowledge manage-ment which offers IE-based facilities for metadata creation, storage and conceptual search. The system has a server-based core that performs ontology-based IE and stores results in a central knowledge base. This server platform can then be used by diverse applications as a service for annotating and querying document spaces.

The ontology-based Information Extraction in KIM produces anno-tations linked both to the ontological class and to the exact individual in the instance base. For new (previously unknown) entities, new identifiers are allocated and assigned; then minimal descriptions are added to the semantic repository. The annotations are kept separately from the content, and an API for their management is provided.

The instance base of KIM has been pre-populated with 200 000 entities of general importance that occur frequently in documents. The majority are different kinds of locations: continents, countries, cities, etc. Each location has geographic co-ordinates and several aliases (usually including English, French, Spanish and sometimes the local transcription of the location name) as well as co-positioning relations (e.g., subRegionOf.).

The difference between TAP and KIM instance base is in the level of ambiguity—TAP has few entities sharing the same alias, while KIM has a lot more, due to its richer collection of locations. Another important difference between KIM and SemTag is their goal. SemTag aims only at accurate classification of the mentions that were found by matching the lexicalisations in the ontology. KIM, on the other hand, is also aiming at finding all mentions, that is coverage, as well as accuracy. The latter is a harder task because there tends to be a trade-off between accuracy and coverage. In addition, SemTag does not attempt to discover and classify new instances, which are not already in the TAP ontology. In other words, KIM performs two tasks—ontology population with new instances and semantic annotation, while SemTag performs only semantic annotation.

3.5.5. KIM Front-ends

KIM has a number of different front-end user interfaces and ones customised for specific applications are easily added. These front-ends provide full access to KIM functionality, including semantic indexing, semantic repositories, metadata annotation services and document and metadata management. Some example front-ends appear below.

The KIM plug-in for Internet Explorer[2] provides lightweight delivery of semantic annotations to the end user. On its first tab, the plug-in displays the ontology and each class has a color used for highlighting the metadata of this type. Classes of interest are selected by the user via check boxes. The user requests the semantic annotation of the currently viewed page by pressing the Annotate button. The KIM server returns the automatically created metadata with its class and instance identifiers. The results are highlighted in the browser window, and are hyperlinked to the KIM Explorer, which displays further information from the ontology about a given instance (see top right window).

The text boxes on the bottom right of Figure 3.3 that contain the *type* and *unique identifier* are seen as tool-tips when the cursor is positioned over a semantically annotated entity.

Selecting the 'Entities' tab of the plug-in generates a list of entities recognised in the current document, sorted by frequency of appearance, as shown in Figure 3.4. This tab also has an icon to execute a semantic

---

[2] KIM Plug-in is available from http://www.ontotext.com/kim

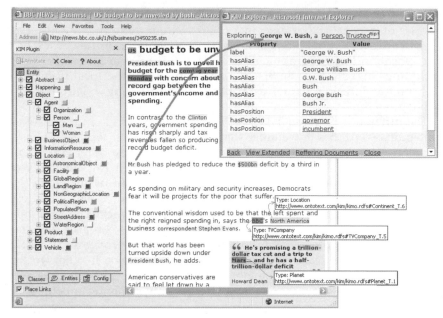

Figure 3.3. KIM plug-in showing the KIM ontology and KB explorer.

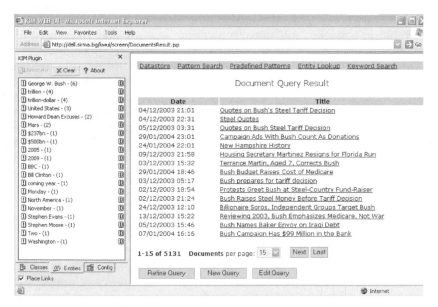

Figure 3.4 KIM plug-in: viewing recognised entities.

query. The result is then shown as a list of documents. The goal is to enable users, while browsing and annotating, to find seamlessly other related documents by selecting one or more entities from the current document.

Another front-end is the KIM Web UI, which offers a powerful semantic search interface. This facet of KIM's capabilities is discussed further in Chapter 8.

3.6. DETERMINISTIC ONTOLOGY AUTHORING USING CONTROLLED LANGUAGE IE

Formalising knowledge in ontologies is a high initial barrier to entry for small organisations and individuals wishing to make data available to semantic knowledge technology, due to the complexity of the standards involved and the high level of experience and engineering skills required by existing ontology authoring environments.

Human language, the most natural method of communication for people, has very complex structures and a large degree of ambiguity. As already discussed in earlier sections, this makes it difficult to process automatically and machines can currently extract a limited amount of the information therein. On the other side of the coin, formal data that is rigidly structured is easily processed by machines but hard and unnatural for people to use. The approach proposed by us bridges that gap by defining a controlled language which, while restricted, still feels natural to people and at the same time is simple enough and unambiguous for the machines to process.

A controlled language is a subset of a natural language which is generally designed to be less ambiguous than the complete language and to include only certain vocabulary terms and grammar rules which are relevant for a specific task. The idea of controlled languages is not new, early controlled languages can trace their roots to 1970s Caterpillar Fundamental English (CFE). The aim there was to restrict the complexity of the language used (CFE only had 850 terms) so that the text is unambiguous enough that it can reliably be translated automatically into a variety of other languages. Further examples are the Caterpillar Technical English (CTE) which had 70 000 carefully chosen domain-specific terms or the KANTOO system developed at Carnegie Mellon University.

Though controlled languages can restrict the colourfulness of expression, they can be used to efficiently communicate concrete information. In most cases using a CL is an exercise in expressing information more consistently and concisely.

In order to facilitate knowledge acquisition and maintenance, we defined a controlled language CLIE CL, modelled to allow maximum expressibility within the smallest set of syntactic structures. The limited number of allowed syntactic sentence structures makes the language easier to learn, much easier to use than OWL, RDF, or SQL for instance. While the syntactic structure of the sentences is constrained, the vocabulary permitted is unrestricted: apart from a small number of key phrases that are used to mark phenomena of interest, any terms can be used freely. This allows for the ontologies created to be open-domain.

The types of actions that are possible are definition of new classes, creation of hierarchies between classes, definition of object and data-type properties, creation of instances and setting of property values for instances.

The greatest advantage of this approach is that it requires essentially no training; there are no complicated user interfaces to be learnt, there are no complex formalisms to be understood. The user can simply start from a simple example which shows all the types of utterances accepted by the system and continue the ontology authoring work by re-using and modifying those examples provided. After the editing is finished, the resulting ontology can then be previewed using a simple ontology viewer implemented for this scope. Once the output has been validated, the ontology can be saved into a variety of formats including RDF(S) and OWL variants.

The language analysis is carried out by an Information Extraction application based on the GATE language processing framework (Cunningham *et al.*, 2002). It comprises some existing GATE components, that is the English tokeniser, part-of-speech tagger and morphological analyser, followed by a cascade of finite-state transducers, based on GATE's JAPE pattern matching language. The role of the transducers is to search for patterns over annotations looking for constructs conforming with the controlled language. In successfully parsed sentences specific tokens are used to extract information.

The tokeniser separates input into tokens. It identifies words within a sentence and separates them from punctuation. For example in the sentence:

```
There are deliverables.
```

The tokens are:

```
[There] [are] [deliverables] [.]
```

The tagger finds the parts of speech for each of the tokens. In other words it finds out what kind of a word each of the tokens is; whether it is a noun, an adjective, a verb, etc.

```
[There]: existential quantifier
[are]: verb—3rd person singular present
[deliverables]: noun—plural
```

The morphological analyser gives the roots of all the words.

```
[There]: root—there
[are]: root—be
[deliverables]: root—deliverable
```

The morphological analyser allows these general types of sentences which announce existence of classes without the need for using artificial singular expressions, that is *There is deliverable*.

The JAPE transducers take the above annotated sentence and look for and mark noun patterns which are likely candidates for CLASS, INSTANCE, and other ontological objects. They also look for specific patterns to extract the information, such as:

```
There are ———.
```

which triggers the creation of one or more new classes in the ontology.

The Controlled Language IE (CLIE) application employs a deterministic approach, so that each sentence can be parsed in one way only. Allowed sentences are unambiguous, so each sentence type is translated to one type of statement. If parsing fails it will result in a warning and no ontological output. In certain cases where the input is invalid, the system will try a less strict analysis mode in order to suggest how such repair may be effected.

For example the text:

There are projects. There are workpackages, tasks and deliverables.

SEKT is a project. MUSING, 'Knowledge Web', and 'Presto Space' are projects.

Projects have workpackages. Workpackages can have tasks. WP1, WP2, WP3, WP4, WP5 and WP6 are workpackages.

SEKT has WP1. MUSING has WP2, WP3 and WP4. 'Knowledge Web' has WP5 and WP6.

generates the ontology shown in Figure 3.5, which is then saved as OWL:

The use of linguistic analysis allows for small variations in the surface form used to name objects (for instance the use of plurals where it feels appropriate from a linguistic point of view) without affecting the capability of the system of identifying different references for the same entity. For example the sentence 'There are animals' will create a new ontology class with the name 'Animal' and the sentence 'Cat is a type of animal' will create a new class with the name 'Cat' as a subclass of the 'Animal' class. The 'Animal' class is referred to in two different ways: one capitalised and in plural form and another lower case and singular. There is also support for listing items; so a sentence like 'There are

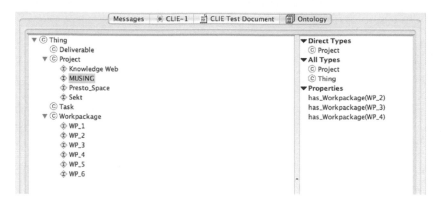

Figure 3.5 Ontology generation from natural language.

projects, work packages and deliverables' will lead to the creation of three new classes: 'Project', 'Work_Package' and 'Deliverable'. The names of entities are normalised—first letters are capitalised, spaces are replaced with underscores and the head word in the case of noun phrases is shown un-inflected. If this is undesirable, names can be included in single quotes which will cause them to be used as they appear in the text.

CLIE can be used in one of two modes—to create a new ontology or to add information to an existing one. Extending an existing ontology requires that names of concepts, instance and properties in the text are first checked against those already in the ontology and only added if necessary. The domain and range restrictions of properties are also checked to ensure consistency.

3.7. CONCLUSION

This chapter motivated the need for Semantic Web enabled Human Language Technology tools and discussed existing systems and out-standing challenges in this area. It introduced the idea of a 'language loop' and showed how HLT can be used to bridge the gap between the current web of language and the Semantic Web.

A number of practical semantic-based applications using HLT-based automatic annotation have already been developed successfully in areas such as market intelligence, financial analysis, media monitoring, etc. For instance, Ontotext have now released a product for recruitment intelligence, which automatically discovers and consolidates jobs vacancies by crawling company web pages. The results from the web mining process are stored in an ontology and conceptual search provides users with information about the latest vacancies by industry sector, region, required skills, etc. In addition, analysts can glean market intelligence from the data by tracking which companies are most active in a given industry sector; what types of expertise they are looking for; and consequently what types of services or products might be of interest to them.

Progress in the development of the Information Society has seen a truly revolutionary decade. Dot com crash notwithstanding, all our lives have been radically changed by the advent of widespread public networking. We believe that a new social revolution is imminent, involving the transition from Information Society to Knowledge Society.

REFERENCES

ACE. 2004. Annotation Guidelines for Entity Detection and Tracking (EDT). Available at http://www.ldc.upenn.edu/Projects/ACE/.

Appelt D. 1999. An introduction to information extraction. *Artificial Intelligence Communications* 12(3):161–172.

Berners-Lee T. 1999. *Weaving the Web*. Orion Business Books, London.

Bowker L. 1995. *A multidimensional approach to classification in Terminology: Working with a computational framework*. PhD thesis, University of Manchester, England.

Cimiano P, Handschuh S, Staab S. 2004. Towards the self-annotating web. In *Proceedings of WWW'04*.

Ciravegna F, Wilks Y. 2003. Designing adaptive information extraction for the semantic web in Amilcare. In Handschuh S, Staab S, (eds) *Annotation for the Semantic Web*, IOS Press, Amsterdam.

Cowie J, Lehnert W. 1996. Information extraction. *Communications of the ACM* 39(1):80–91.

Cunningham H. 2005. *Information Extraction, Automatic Encyclopedia of Language and Linguistics* (2nd edn).

Cunningham H, Maynard D, Bontcheva K, Tablan V. 2002. GATE: A Framework and Graphical Development Environment for Robust NLP Tools and Applications. In Proceedings of the 40th Anniversary Meeting of the Association for Computational Linguistics (ACL'02).

Davies J, Fensel D, van Harmelen F (Eds). 2002. *Towards the Semantic Web: Ontology-Driven Knowledge Management*. John Wiley & Sons, Ltd: New York.

Dean M, Schreiber G, Bechhofer S, van Harmelen F, Hendler J, Horrocks I, McGuinness D L, Patel-Schneider P F, Stein L A. 2004. OWL web ontology language reference: W3C recommendation, *W3C*, February, available at: www.w3.org/TR/owl-ref/.

Dill S, Eiron N, Gibson D, Gruhl D, Guha R, Jhingran A, Kanungo T, Rajagopalan S, Tomkins A, Tomlin JA, Zien JY. 2003. SemTag and Seeker: Bootstrapping the semantic web via automated semantic annotation. In *Proceedings of WWW' 03*.

Domingue J, Dzbor M, Motta E. 2004. Magpie: Supporting Browsing and Navigation on the Semantic Web. In Nunes N, Rich C, (eds). *Proceedings ACM Conference on Intelligent User Interfaces (IUI)*, pp 191–197.

Felber H. 1984. *Terminology Manual*. Unesco and Infoterm, Paris.

Fensel D, Hendler J, Wahlster W, Lieberman H (Eds). 2002. *Spinning the Semantic Web: Bringing the World Wide Web to Its Full Potential*. MIT Press: Cambridge, MA.

Handschuh S, Staab S, Ciravegna F. 2002. S-CREAM—Semi-automatic CREAtion of Metadata. In *13th International Conference on Knowledge Engineering and Knowledge Management (EKAW02)*, Siguenza, Spain, pp 358–372.

Handschuh S, Staab S, Maedche A. 2001. CREAM—Creating relational metadata with a component-based, ontology-driven framework. In *Proceedings of K-CAP 2001*, Victoria, BC, Canada.

Humphreys K, Gaizauskas R, Azzam S, Huyck C, Mitchell B, Cunningham H, Wilks Y. 1998. Description of the LaSIE system as used for MUC-7. In *Proceedings of the Seventh Message Understanding Conference (MUC-7)*. http://www.itl.nist.gov/iaui/894.02/-related projects/muc/index.html.

Kiryakov A, Popov B, Terziev I, Manov D, Ognyanoff D. 2005. Semantic annotation, indexing and retrieval. *Journal of Web Semantics* 2(1).

Kogut P, Holmes W. 2001. AeroDAML: Applying Information Extraction to Generate DAML Annotations from Web Pages. In *First International Conference on Knowledge Capture (K-CAP 2001), Workshop on Knowledge Markup and SemanticAnnotation*, Victoria, B.C.

Mahesh K, Kud J, Dixon P. 1999. *Oracle at TREC8: A Lexical Approach*. In Proceeding of the Eighth Text Retrieval Conference (TREC-8).

Mahesh K, Nirenburg S, Cowie J, Farwell D. 1996. An Assessment of Cyc for Natural Language Processing. Technical Report MCCS Report, New Mexico State University.

Moldovan D, Mihalcea R. 2001. *Document Indexing Using Named Entities*. In Studies in Informatics and Control, Vol. 10, No. 1.

Motta E, Vargas-Vera M, Domingue J, Lanzoni M, Stutt A, Ciravegna F. 2002. MnM: Ontology driven semi-automatic and automatic support for semantic markup. In *13th International Conference on Knowledge Engineering and Knowledge Management (EKAW02)*, Siguenza, Spain, pp 379–391.

Pazienza M T (ed.). 2003. *Information Extraction in the Web Era*. Springer-Verlag: New York.

Pustejovsky J, Boguraev B, Verhagen M, Buitelaar P, Johnston M. 1997. *Semantic Indexing and Typed Hyperlinking*. In Proceedings of the AAAI Conference, Spring Symposium, NLP for WWW, Stanford University, CA, pp 120–128.

SAIC. 1998. Proceedings of the Seventh Message Understanding Conference (MUC-7), http://www.itl.nist.gov/iaui/894.02/related projects/muc/index.html.

Voorhees E. 1998. Using WordNet for text retrieval. In *WordNet: An Electronic Lexical Database*, Fellbaum C (ed.). MIT Press.

4

Ontology Evolution

Stephan Bloehdorn, Peter Haase, York Sure and Johanna Voelker

4.1. INTRODUCTION

In our knowledge-intensive economy, the amount of available knowledge stored, for example, in digital libraries and other knowledge repositories, increases ever more rapidly, as does our reliance on being able to locate and exploit relevant information. Knowledge workers rely heavily on the availability and accessibility of knowledge contained in such repositories. The sheer mass of knowledge available today, however, requires sophisticated support for searching and, often considered as equally important, personalization.

Ontology and metadata technology is one approach for addressing such challenges (Davies *et al.*, 2005). Ontologies (Staab and Studer, 2004) enable knowledge to be made explicit, formalise the relevant underlying view of the world (domain model) and make such models machine processable and interpretable. The use of ontologies and associated metadata offers the prospect of significant improvement to the information retrieval task, as discussed in more detail in Chapter 8. The classification of documents according to a given topic hierarchy facilitates structuring and browsing of huge document collections; semantic annotation of individual documents improves the precision of search queries or even allows for sophisticated question answering; and semantic user profiles representing the current working context of the user can be used to guide searching, browsing, and alerting.

Ontologies, to be effective, need to change as fast as the parts of the world they describe. There are two main challenges in adapting ontologies.

Semantic Web Technologies: Trends and Research in Ontology-based Systems
John Davies, Rudi Studer, Paul Warren   © 2006 John Wiley & Sons, Ltd

The evolution of ontologies should reflect both the *changing interests of people* and the *changing data*, for example the documents stored in a digital library. In this chapter, we present an overview of the state-of-the-art in ontology evolution with a special focus on change discovery for ontologies. We would like to mention that our approach supports specific steps of the DILIGENT methodology for ontology engineering, described in Chapter 9.

In this work, we will distinguish *change capturing* and *change discovery*. The task of *change capturing* can be defined as the *generation of ontology changes* from explicit and implicit requirements. Explicit requirements are generated, for example, by ontology engineers who want to adapt the ontology to new requirements or by the end-users who provide explicit feedback about the usability of ontology entities. We call the changes resulting from this kind of requirements *top-down changes*. Implicit requirements leading to so-called *bottom-up changes* are reflected in the behavior of the system and can be induced by means of *change discovery* methods. While usage-driven changes arise out of usage patterns of the ontology, data-driven changes are generated by modifications of the reference-data such as text documents or a database which contains the knowledge modeled by the ontology.

The remainder of this chapter is structured as follows. In Section 2, we present an overview of the state-of-the-art in ontology evolution. In Section 3, we present a logical architecture for ontology evolution, exemplified in the context of a digital library. The main components of this logical architecture are then described in detail. In Sections 4 and 5, we illustrate techniques that deal with usage-driven ontology changes and data-driven ontology changes, respectively. In the former approach, changes are recommended based on the actual usage of the ontologies; in the latter approach we make use of the constant flows of documents coming into, for example a digital library to keep ontologies up-to-date. Finally, we conclude in Section 6.

4.2. ONTOLOGY EVOLUTION: STATE-OF-THE-ART

In this section, we provide an overview of the state-of-the-art in ontology evolution. In Stojanovic *et al.* (2002), the authors identify a possible six-phase evolution process (as shown in Figure 4.1), the phases being:

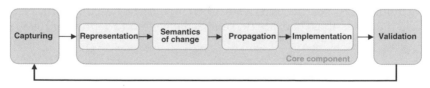

Figure 4.1 Ontology evolution process.

(1) change capturing, (2) change representation, (3) semantics of change, (4) change implementation, (5) change propagation, and (6) change validation. In the following, we will use this evolution process as the basis for an analysis of the state-of-the-art.

4.2.1. Change Capturing

The process of ontology evolution starts with capturing changes either from explicit requirements or from the result of *change discovery* methods, which induce changes from patterns in data and usage. Explicit requirements are generated, for example, by ontology engineers who want to adapt the ontology to new requirements or by the end-users who provide the explicit feedback about the usability of ontology entities. The changes resulting from such requirements are called *top-down* changes. Implicit requirements leading to so-called *bottom-up* changes are reflected in the behavior of the system and can be discovered only through the analysis of this behavior. Stojanovic (2004) defines different types of change discovery, we put in this work a focus on usage-driven and data-driven change discovery.

Usage-driven changes result from the usage patterns created over a period of time. Once ontologies reach certain levels of size and complexity, the decision about which parts remain relevant and which are outdated is a huge task for ontology engineers. Usage patterns of ontologies and their metadata allow the detection of often or less often used parts, thus reflecting the interests of users in parts of ontologies. They can be derived by tracking querying and browsing behaviors of users during the application of ontologies as shown in Stojanovic *et al.* (2003b).

Stojanovic (2004) defines data-driven change discovery as the problem of deriving ontological changes from the ontology instances by applying techniques such as data-mining, Formal Concept Analysis (FCA) or various heuristics. For example, one possible heuristic might be: if no instance of a concept C uses any of the properties defined for C, but only properties inherited from the parent concept, C is not necessary. An implementation of this notion of data-driven change discovery is included in the KAON tool suite (Maedche *et al.*, 2003).

Here we use a more general definition of data-driven change discovery based on the assumption that an ontology is often learned or constructed in order to reflect the knowledge more or less implicitly given by a number of documents or a database. Therefore, any change to the underlying data set, such as a newly added document or a changed database entry, might require an update of the ontology. Data-driven change discovery can be defined as the task of deriving ontology changes from modifications to the knowledge from which the ontology has been constructed. One difference between these two definitions is that the latter always assumes an existing ontology, while the former can be applied to an empty ontology as well, but requires an evolving data set associated with this ontology.

Ontology engineering follows well-established processes such as described by Sure *et al.* (2002a). So far, one has distinguished between manual and (semi-)automatic approaches to ontology engineering. If the ontology creation process is done manually, for example by a knowledge engineer in collaboration with domain experts supported by an ontology engineering system such as OntoEdit (Sure *et al.*, 2002b), then both general and concrete relationships need to be held in the mind of this knowledge engineer. This requires a significant manual effort for codifying knowledge into ontologies. On the other hand, if the process of creating the ontology is done semi- or fully automatically with the help of an ontology learning system such as Text2Onto (Cimiano and Völker, 2005) these general and concrete relationships are generated and represented explicitly by the system. Of course, the first kind of knowledge is always given by the specific implementation of the ontology learning algorithms which are used. However, in order to enable an existing ontology learning system to support data-driven change discovery, it is necessary to make it store all available knowledge about concrete relationships between ontology entities and the data set.

4.2.2. Change Representation

To resolve changes, they have to be identified and represented in a suitable format which means that the change representation needs to be defined for a given ontology model. Changes can be represented on various levels of granularity, for example as elementary or complex changes.

The set of ontology change operations depends heavily on the underlying ontology model. Most existing work on ontology evolution builds on frame-like or object models, centred around classes, properties, etc.

Stojanovic (2004) derives a set of ontology changes for the KAON ontology model. The author specifies fine-grained changes that can be performed in the course of the ontology evolution. They are called elementary changes, since they cannot be decomposed into simpler changes. An elementary change is either an *add* or *remove* transformation, applied to an entity in the ontology model. The author also mentions that this level of change representation is not always appropriate and therefore introduces the notion of composite changes: a composite change is an ontology change that modifies (creates, removes or changes) one and only one level of neighborhood of entities in the ontology, where the neighborhood is defined via structural links between entities. Examples for such composite changes would be: 'Pull concept up,' 'Copy Concept,' 'Split Concept,' etc. Further, the author introduces complex changes: a complex change is an ontology change that can be decomposed into any combination of at least two elementary or composite ontology changes. As a result, the author places the identified types of changes into a taxonomy of changes.

Klein and Noy (2003) also state that information about changes can be represented in many different ways. They describe different representations and propose a framework that integrates them. They show how different representations in the framework are related by describing some techniques and heuristics that supplement information in one representation with information from other representations and present an ontology of change operations, which is the kernel of the framework. Klein (2004) describes a set of changes for the OWL ontology language, based on an OWL meta-model. Unlike the previously mentioned set of KAON ontology changes, the author considers also *Modify* operations in addition to *Delete* and *Add* operations. Further, the taxonomy contains *Set* and *Unset* operations for properties (e.g., to set transitivity). The author introduces an extensive terminology of change operations along two dimensions: *atomic* versus *composite* and *simple* versus *rich*. *Atomic operations* are operations that cannot be subdivided into smaller operations, whereas *composite operations* provide a mechanism for grouping operations that constitute a logical entity. *Simple changes* can be detected by analyzing the structure of the ontology only, whereas *rich changes* incorporate information about the implication of the operation on the logical model of the ontology, for their identification one thus needs to query the logical theory of the ontology. The author also proposes a method for finding complex ontology changes. It is based on a set of rules and heuristics to generate a complex change from a set of basic changes. Both Stojanovic (2004) and Klein (2004) present an 'ontology for ontology changes' for their respective ontology language and identified change operations.

Another form of change representation for OWL is defined by Haase and Stojanovic (2005), who follow an ontology model influenced by Description Logics, which treats an ontology as a knowledge base consisting of a set of axioms. Accordingly, they allow the atomic change operations of adding and removing axioms. Obviously, representing changes at the level of axioms is very fine grained. However, based on this minimal set of atomic change operations, it is possible to define more complex, higher-level descriptions of ontology changes. Composite ontology change operations can be expressed as a sequence of atomic ontology change operations. The semantics of the sequence is the chaining of the corresponding functions.

Models for change representations for other ontology languages exist, too: a formal method for tracking changes in the RDF repository is proposed in Ognyanov and Kiryakov (2002). The RDF statements are pieces of knowledge they operate on. The authors argue that during ontology evolution, the RDF statements can be only deleted or added, but not changed. Higher levels of abstraction of ontology changes such as composite and complex ontology changes are not considered at all in that approach.

4.2.3. Semantics of Change

The ontology change operations need to be managed such that the ontology remains consistent throughout. The consistency of an ontology is defined in terms of consistency conditions, or invariants that must be satisfied by the ontology. The meaning of consistency depends heavily on the underlying ontology model. It can for example be defined using a set of constraints or it can be given a model-theoretic definition. In the following we provide an overview of various notions of consistency and approaches for the realization of the changes.

Consistency: Stojanovic (2004) defines consistency as: 'An ontology is defined to be consistent with respect to its model if and only if it preserves the constraints defined for the underlying ontology model.'

For example, in the KAON ontology model, the consistency of ontologies is defined using a set of constraints, called invariants. These invariants state for example that the concept hierarchy has to be a directed acyclic graph.

In Haase and Stojanovic (2005), the authors describe the semantics of change for the consistent evolution of OWL ontologies, considering the *structural*, *logical*, and *user-defined* consistency conditions:

- *Structural Consistency* ensures that the ontology obeys the constraints of the ontology language with respect to how the constructs of the ontology language are used.
- *Logical Consistency* regards the formal semantics of the ontology: viewing the ontology as a logical theory, an ontology as logically consistent if it is satisfiable, meaning that it does not contain contradicting information.
- *User-defined Consistency*: Finally, there may be definitions of consistency that are not captured by the underlying ontology language itself, but rather given by some application or usage context. The conditions are explicitly defined by the user and they must be met in order for the ontology to be considered consistent.

Stojanovic (2004) describes and compares two approaches to verify ontology consistency:

1. *a posteriori verification*, where first the changes are executed, and then the updated ontology is checked to determine whether it satisfies the consistency constraints.
2. *a priori verification*, which defines a respective set of preconditions for each change. It must be proven that, for each change, the consistency

will be maintained if (1) an ontology is consistent prior to an update and (2) the preconditions are satisfied.

Realization: Stojanovic *et al.* (2002, 2003a) describe two approaches for the realization of the semantics of change, a procedural and a declarative one, respectively. In both these approaches, the KAON ontology model is assumed. The two approaches were adopted from the database community and followed to ensure ontological consistency (Franconi *et al.*, 2000):

1. *Procedural approach*: this approach is based on the constraints, which define the consistency of a schema, and definite rules, which must be followed to maintain constraints satisfied after each change.
2. *Declarative approach*: this approach is based on the sound and complete set of axioms (provided with an inference mechanism) that formalises the dynamics of the evolution.

In Stojanovic *et al.* (2003a) (declarative approach), the authors present an approach to model ontology evolution as reconfiguration-design problem solving. The problem is reduced to a graph search where the nodes are evolving ontologies and the edges represent the changes that transform the source node into the target node. The search is guided by the constraints provided partially by the user and partially by a set of rules defining ontology consistency. In this way they allow a user to specify an arbitrary request declaratively and ensure its resolution.

In Stojanovic *et al.* (2002) (procedural approach), the authors focus on providing the user with capabilities to control and customize the realization of the semantics of change. They introduce the concept of an evolution strategy encapsulating policy for evolution with respect to the user's requirements. To resolve a change, the evolution process needs to determine answers at many *resolution points*—branch points during change resolution were taking a different path will produce different results. Each possible answer at each resolution point is an *elementary evolution strategy*. A common policy consisting of a set of elementary evolution strategies—each giving an answer for one resolution point—is an *evolution strategy* and is used to customize the ontology evolution process. Thus, an evolution strategy unambiguously defines the way elementary changes will be resolved. Typically a particular evolution strategy is chosen by the user at the start of the ontology evolution process.

A similar approach is followed by Haase and Stojanovic (2005) for the consistent evolution of OWL ontologies: here resolution strategies map each consistency condition to a resolution function, which returns for a given ontology and an ontology change operation an additional change operation. Further it is required that for all possible ontologies and for all possible change operations, the assigned resolution function generates changes, which—applied to the ontology—result in an ontology that satisfies the consistency condition.

The semantics of OWL ontologies is defined via a model theory, which explicates the relationship between the language syntax and the model of a domain: an interpretation satisfies an ontology, if it satisfies each axiom in the ontology. Axioms thus result in semantic conditions on the interpretations. Consequently, contradictory axioms will allow no possible interpretations. Please note that because of the monotonicity of the logic, an ontology can only become inconsistent by adding axioms: if a set of axioms is satisfiable, it will still be satisfiable when any axiom is deleted. Therefore, the consistency only needs to be checked for ontology change operations that add axioms to the ontology.

The goal of the resolution function is to determine a set of axioms to be removed, in order to obtain a logically consistent ontology with 'minimal impact' on the existing ontology. Obviously, the definition of minimal impact may depend on the particular user requirements. A very simple definition could be that the number of axioms to be removed should be minimized. More advanced definitions could include a notion of confidence or relevance of the axioms. Based on this notion of 'minimal impact' we can define an algorithm that generates a minimal number of changes that result in a *maximally consistent subontology*, that is a subontology to which no axiom from the original ontology can be added without losing consistency.

In many cases it will not be feasible to resolve logical inconsistencies in a fully automated manner. In this case, an alternative approach for resolving inconsistencies allows the interaction of the user to determine which changes should be generated. Unlike the first approach, this approach tries to localize the inconsistencies by determining a minimal inconsistent subontology, which intuitively is a minimal set of contradicting axioms. Once we have localized this minimal set, we present it to the user. Typically, this set is considerably smaller than the entire ontology, so that it will be easier for the user to decide how to resolve the inconsistency. Algorithms to find maximally consistent and minimally inconsistent subontologies based on the notion of a *selection function* are described in Haase and Stojanovic (2005).

Finally, it should be noted that there exist other approaches to deal with inconsistencies, for example, Haase *et al.* (2005) compare consistent evolution of OWL ontologies with other approaches in a framework for dealing with inconsistencies in changing ontologies.

4.2.4. Change Propagation

Ontologies often reuse and extend other ontologies. Therefore, an ontology update might potentially corrupt ontologies depending (through inclusion, mapping integration, etc.) on the modified ontology and consequently, all the artefacts based on these ontologies. The task of the change propagation phase of the ontology evolution process is to ensure consistency of *dependent artefacts* after an ontology update has been performed. These artefacts may include dependent ontologies, instances, as well as application programs using the ontology.

Maedche *et al.* (2003) present an approach for evolution in the context of dependent and distributed ontologies. The authors define the notion of *Dependent Ontology Consistency*: a dependent ontology is consistent if the ontology itself and all its included ontologies, observed alone and independently of the ontologies in which they are reused, are *single ontology consistent*. *Push-based* and *Pull-based* approaches for the synchronization of dependent ontologies are compared. The authors follow a push-based approach for dependent ontologies on one node (nondistributed) and present an algorithm for dependent ontology evolution.

Further, for the case of multiple ontologies distributed over multiple nodes, Maedche *et al.* (2003) define *Replication Ontology Consistency* [an ontology is replication consistent if it is equivalent to its original and all its included ontologies (directly and indirectly) are replication consistent]. For the synchronization between originals and replicas, they follow a pull-based approach.

4.2.5. Change Implementation

The role of the change implementation phase of the ontology evolution process is (i) to inform an ontology engineer about all consequences of a change request, (ii) to apply all the (required and derived) changes, and (iii) to keep track of performed changes.

Change Notification: In order to avoid performing undesired changes, a list of all implications for the ontology and dependent artefacts should be generated and presented to the ontology engineer, who should then be able to accept or abort these changes.

Change Application: The application of a change should have transactional properties, that is (A) Atomicity, (C) Consistency, (I) Isolation, and (D) Durability. The approach of Stojanovic (2004) realizes this requirement by the strict separation between the request specification and the change implementation. This allows the set of change operations to be easily treated as one atomic transaction, since all the changes are applied at once.

Change Logging: There are various ways to keep track of the performed changes. Stojanovic (2004) proposes an *evolution log* based on an *evolution ontology* for the KAON ontology model. The evolution ontology covers the various types of changes, dependencies between changes (causal dependencies as well as ordering), as well as the decision-making process.

4.2.6. Change Validation

There are numerous circumstances where it can be desirable to reverse the effects of the ontology evolution, as for example in the following cases:

- The ontology engineer may fail to understand the actual effect of the change and approve a change which should not be performed.
- It may be desired to change the ontology for experimental purposes.
- When working on an ontology collaboratively, different ontology engineers may have different ideas about how the ontology should be changed.

It is the task of the change validation phase to recover from these situations. Change validation enables justification of performed changes or undoing them at user's request. Consequently, the usability of the ontology evolution system is increased.

4.3. LOGICAL ARCHITECTURE

In this section, we present a logical architecture tailored to support the evolution of ontologies in a digital library or other electronic information repositories. Figure 4.2 illustrates the connections between the components of the overall architecture.

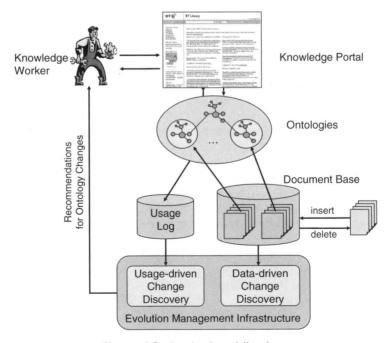

Figure 4.2 Logical architecture.

In this architecture, a knowledge worker interacts with a knowledge portal to access the content of the digital library, which comprises several document databases, organized using ontologies. The interaction is recorded in a usage log. This usage information and the information about changes in the document base are exploited to recommend changes to the ontologies, thus closing the loop with the knowledge worker.

Knowledge Worker: The knowledge worker primarily consumes knowledge from the digital library. He uses the digital library to fulfill a particular information need. However, a knowledge worker may also contribute to the digital library, either by contributing content or by organizing the existing content, providing metadata, etc. In particular, a knowledge worker can take the role of an ontology engineer.

Knowledge Portal: The knowledge worker interacts with the knowledge portal as the user interface. It allows the user to search the library's contents, and it presents the contents in an organized way. The knowledge portal may also provide the knowledge worker with information in a proactive manner, for example by notification, etc.

Document Base: The document base comprises a corpus of documents. In the context of the digital library, these documents are typically text documents, but may also include multimedia content such as audio, video, and images. While we treat the document as one logical unit, it may actually consist of a number of distributed sources. The content of the document base typically is not static, but changes over time: new documents come in, but also documents may be removed from the document base.

Ontologies: Ontologies are the basis for rich, semantic descriptions of the content in the digital library. Here, we can identify two main modules of the ontology: the *application ontology* describes different generic aspects of bibliographic metadata (such as author, creation data) and are valid across various bibliographic sources. *Domain ontologies* describe aspects that are specific to particular domains and are used as a conceptual backbone for structuring the domain information. Such a domain ontology typically comprises conceptual relations, such as a topic hierarchy, but also richer taxonomic and nontaxonomic relations.

While the application ontology can be assumed to be fairly static, the domain ontologies must be continuously adapted to the changing needs. The ontologies are used for various purposes: first of all, the documents in the document base are annotated and classified according to the ontology. This ontological metadata can then be exploited for advanced knowledge access, including navigation, browsing, and semantic searches. Finally, the ontology can be used for the visualization of results, for example for displaying the relationships between information objects.

Usage Log: The interaction of the knowledge worker with the knowledge portal is recorded in a usage log. Of particular interest is how the ontology has been used in the interaction, that is which elements have been queried, which paths have been navigated, etc. By tracking the users' interactions with the application in a log file, it is possible to collect useful information that can be used to assess the main interests of the users. In this way, we are able to obtain implicit feedback and to extract ontology change requirements to improve the interaction with the application.

Evolution Management: The process of ontology evolution is supported by the evolution management infrastructure. The first important aspect is the discovery of changes. While in some cases changes to the ontology may be requested explicitly, the actual challenge is to obtain and to examine the nonexplicit but available knowledge about the needs of the end-users. This can be done by analyzing various data sources related to the content that is described using the ontology. It can also be done by analyzing the end-user's behavior which leads to information about her likes, dislikes, preferences or the way she behaves. Based on the analysis of this information, suggested ontology changes can be made to the knowledge worker. This results in an ontology better suited to the needs of end-users. In the following sections, we will discuss the possibility of continuous ontology improvement by semi-automatic discovery of such changes, that is data-driven and usage-driven ontology evolution.

4.4. DATA-DRIVEN ONTOLOGY CHANGES

Since many real-world data sets tend to be highly dynamic, ontology management systems have to deal with potential inconsistencies between the knowledge modeled by ontologies and the knowledge given by the underlying data. Data-driven change discovery targets this problem by providing methods for automatic or semi-automatic adaptation of ontologies according to modifications being applied to the underlying data set.

Suppose, for example, a user wants to find information about the SEKT project. When searching for SEKT (as a search string) with a typical search engine he will probably find a lot of pages, mostly about sparkling wine (since this is the most common meaning of the word SEKT in German), which are not relevant with respect to his actual information need. Given a more sophisticated semantically enhanced search engine he would have several ways of specifying the semantics of what he wants to find:

- *Ontology-based searching:* The user selects the concept Project from a domain ontology which might have been manually constructed or (semi-)automatically learned from the document base. Then he searches for SEKT as an instance of that concept. The search engine examines the ontological metadata which has previously been added to the content of each document in order to find those documents which are most likely to be relevant to his query.

- *Topic hierarchy/browsing:* Suppose a hierarchy of topics, one of which is *The SEKT project*, is used to classify a corpus of documents. The classification of the documents could, for example, have been done automatically based on ontological knowledge extracted from the documents. The user can choose the topic in which he is interested, in this case *The SEKT Project*, from the topic hierarchy.
- *Contextualized search*: The user simply searches for SEKT and the system concludes from his semantic user profile and his current working context that he is looking for information about a certain (research) project.

Of course, having found some relevant documents the user's information need is not yet satisfied completely, but the number of documents he has to read to find the relevant information about the SEKT project has decreased significantly. Nevertheless, depending on his query and the size of the document base some hundreds of documents might be left. Ontology learning algorithms can be used to provide the user with an aggregated view of the knowledge contained in these documents, showing the user the concepts, instances and relations which were extracted from the text. For this purpose a number of tools such as Text2Onto (Cimiano and Völker, 2005) are available which apply natural language processing as well as machine learning techniques in order to build ontologies in an automatic or semi-automatic fashion. Consider the following example:

PROTON is a flexible, lightweight upper level ontology that is easy to adopt and extend for the purposes of the tools and applications developed within [the] SEKT project (SEKT Deliverable D1.8.1).

From the text fragment cited above you can conclude that SEKT is an instance of the concept *project*. It also tells you that *PROTON* is an instance of *upper-level ontology*, which in turn is a special kind of *ontology*. But such an ontology cannot only be used for browsing. It might also serve as a basis for document classification, metadata generation, ontology-based searching, and the construction of a semantic user profile. All of these applications require a tight relationship between the ontology and the underlying data, that is the ontology must explicitly represent the knowledge which is more or less implicitly given by the document base. Therefore changes to the data should be immediately reflected by the ontology.

Suppose now that the document base is extended, for example by focussed crawling, the inclusion of knowledge stored on the user's desktop or Peer-to-Peer techniques. In this case all ontologies which are affected by these changes have to be adapted in order to reflect the knowledge gained through the additional information available.

Moreover, the ontological metadata associated with each document has to be updated. Otherwise searching and browsing the document base might lead to incomplete or even incorrect results.

Imagine, for example, that the following text fragments are added to a document base consisting of the document cited in the previous example plus a few other documents, which are not about the SEKT project.

Collaboration within SEKT will be enhanced through a programme of joint activities with other integrated projects in the semantically enabled knowledge systems strategic objective (···) *(SEKT Contract Documentation) EU-IST Integrated Project (IP) IST-2003-506826 SEKT (SEKT Deliverable D4.2.1).*

From these two text fragments ontology learning algorithms can extract a previously unknown concept *integrated project* which is a subclass of *project* and which has the same meaning as *IP* in this domain. Furthermore, SEKT will be reclassified as an instance of the concept *integrated project*.

If the user had searched for SEKT as an instance of *IP* before the above-mentioned changes to the document base had been made, there would have been no results. Without the information given by the two newly added documents the system either does not know the concept *IP* or it assumes it to be equivalent to *internet protocol* since the term *IP* is most often used in this sense.

But how can we make sure that all ontologies, as well as dependent annotations and metadata, stay always up-to-date with the document base? One possibility would be a complete re-engineering of the ontology each time the document base changes. But of course, building an ontology for a huge amount of data is a difficult and time-consuming task even if it is supported by tools for automatic or semi-automatic ontology extraction. A much more efficient way would be to adapt the ontology according to the changes, that is to identify for each change all concepts, instances, and relations in the ontology which are affected by this change, and to modify the ontology accordingly.

Therefore, data-driven change discovery aims at providing methods for automatic or semi-automatic adaptation of an ontology, as the underlying data changes.

4.4.1. Incremental Ontology Learning

Independently from a particular use case scenario, the following general prerequisites must be fulfilled by any application, designed to support data-driven change discovery. The most important requirement is, of course the need to keep track of all changes to the data. Each change must be represented in a way which allows it to be associated with various kinds of information, such as its type, the source it has been created from and its target object (e.g., a text document). In order to make the whole system as transparent as possible not only changes to the data set, but also changes to the ontology should be logged. Moreover, if ontological changes are caused by changes to the underlying data, then the ontological changes should be associated with information about the corresponding changes to the data.

Optionally, in order to take different user preferences into account, various change strategies could be defined. This allows the specification of the extent to which changes to the data should change the ontology. For example, a user might want the ontology to be updated in case of newly added or modified data, but, on the other hand, he might want the ontology to remain unchanged if some part of the data set is deleted.

In addition to the above-mentioned requirements, different kinds of knowledge have to be generated or represented within a change discovery system:

1. *Generic* knowledge about relationships between data and ontology is required, since in case of newly added or modified data, additional knowledge has to be extracted and represented by the ontology. For example, generic knowledge may include heuristics of how to identify concepts and their taxonomic relationships in the data.
2. *Concrete* knowledge about relationships between the data and ontology concepts, instances and relations is needed because deleting or modifying information in the data set might have an impact on existing elements in the ontology. This impact has to be determined by the application to generate appropriate ontology changes. The actual references to ontology elements in the data are an example for concrete knowledge.

It is quite obvious that automatic or semi-automatic data-driven change discovery requires a formal, explicit representation of both kinds of knowledge. Since this representation is usually unavailable in case of a manually built ontology, we can conclude that an implementation of data-driven change discovery methods should be embedded in the context of an ontology extraction system. Such systems usually represent general knowledge about the relationship between an ontology and the underlying data set by means of ontology learning algorithms. Consequently, the concrete knowledge to be stored by an ontology extraction system depends on the way these algorithms are implemented. A concept extraction algorithm, for example, might need to store the text references and term frequencies associated with each concept, whereas a pattern-based concept classification algorithm might have to remember the occurrences of all hyponymy patterns matched in the text. Whereas existing tools such as TextToOnto (Mädche and Volz, 2001) mostly neglect this kind of concrete knowledge and therefore do not provide any support for data-driven change discovery, the next generation of ontology extraction systems, including for example Text2Onto (Cimiano and Völker, 2005), will explicitly target the problem of incremental ontology learning.

4.5. USAGE-DRIVEN ONTOLOGY CHANGES

In this section, we will describe how information on the usage of ontologies can be analyzed to recommend changes to the ontology. The usage analysis that leads to the recommendation of changes is a very complex activity. First, it is difficult to find meaningful usage patterns. For example, is it useful for an application to discover that many more users are interested in the topic *industrial project* than in the topic *research*? Second, when a meaningful usage pattern is found, the open issue is how to translate it into a change that leads to the improvement of an application. For example, how to interpret the information that a lot of users are interested in *industrial research project* and *basic research project*, but none of them are interested in the third type of project—*applied research project*.

Since in an ontology-based application, the ontology serves as a conceptual model of the domain, the interpretation of these usage patterns on the level of the ontology alleviates the process of discovering useful changes in the application. The first pattern mentioned above can be treated as useless for discovering changes if there is no relation between the concepts *industrial project* and *research* in the underlying ontology. Moreover, the structure of the ontology can be used as the background knowledge for generating useful changes. For example, in the case that *industrial project*, *basic research project*, and *applied research project* are three sub-concepts of the concept *project* in the domain ontology, in order to tailor the concepts to the users' needs, the second pattern mentioned could lead to either deleting the 'unused' concept *applied research project* or its merging with one of the two other concepts (i.e., industrial research or basic research). Such an interpretation requires the familiarity with the ontology model definition, the ontology itself, as well as experience in modifying ontologies. Moreover, the increasing complexity of ontologies demands a correspondingly larger human effort for its management. It is clear that manual effort can be time consuming and error prone. Finally, this process requires highly skilled personnel, which makes it costly.

The focal point of the approach is the continual adaptation of the ontology to the users' needs. As illustrated above, by analyzing the usage data with respect to the ontology, more meaningful changes can be discovered. Moreover, since the content and layout (structure) of an ontology-based application are based on the underlying ontology, by changing the ontology according to the users' needs, the application itself is tailored to these needs.

4.5.1. Usage-driven Hierarchy Pruning

Our goal is to help an ontology engineer in the continual improvement of the ontology. This support can be split into two phases:

1. To help the ontology engineer find the changes that should be performed; and
2. To help her in performing such changes.

The first phase is focused on discovering some anomalies in the ontology design, the repair of which improves the usability of the ontology. It results in a set of ontology changes. One important problem we face in developing an ontology is the creation of a hierarchy of concepts, since a hierarchy, depending on the users' needs, can be defined from various points of view and on different levels of granularity. Moreover, the users' needs can change over time, and the hierarchy should reflect such a migration. The usage of the hierarchy is the best way to estimate how a hierarchy corresponds to the needs of the users. Consider the example shown in Figure 4.3 (taken from Stojanovic *et al.*, 2003a):

Let us assume that in the initial hierarchy (developed by using one of the above-mentioned approaches), the concept X has ten sub-concepts (c1, c2, ···, c10), that is an ontology engineer has found that these ten concepts correspond to the users' needs in the best way. However, the usage of this hierarchy in a longer period of time showed that about 95 % of the users are interested in just three sub-concepts of these ten. This means that 95 % of the users, as they browse the hierarchy, find 70 % of the sub-concepts irrelevant. Consequently, these 95 % of users invest more time in performing a task than needed, since irrelevant information receives their attention. Moreover, there are more chances to make an accidental error (e.g., an accidental click on the wrong link), since the probability of selecting irrelevant information is bigger.

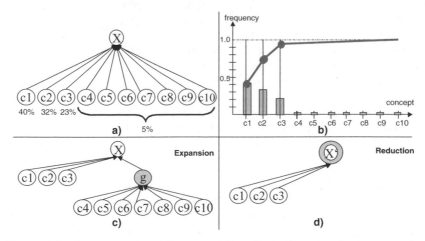

Figure 4.3 An example of the nonuniformity in the usage of concepts.

In order to make this hierarchy more suitable to users' needs, two ways of 'restructuring' the initial hierarchy would be useful:

1. *Expansion*: to move all seven 'irrelevant' subconcepts down in the hierarchy by grouping them under a new sub-concept g (see Figure 4.3(c)).
2. *Reduction*: to remove all seven 'irrelevant' concepts, while redistributing their instances into the remaining sub-concepts or the parent concept (see Figure 4.3(d)).

Through the expansion, the needs of the 5 % of the users are preserved by the newly introduced concept and the remaining 95 % of the users benefit from the more compact structure. By the reduction, the new structure corresponds completely to the needs of 95 % of the users. Moreover, the usability of the ontology has increased, since the instances which were hidden in the 'irrelevant' sub-concepts are now visible for the additional 95 % of the users. Consequently, these users might find them useful, although in the initial classification they are *a priori* considered as irrelevant (i.e., these instances were not considered at all). Note that the Pareto diagram shown in Figure 4.3(b) enables the automatic discovery of the minimal subset of the sub-concepts, which covers the needs of most of the users. For a formalization of this discovery process, including an evaluation study, we refer the interested reader to Stojanovic *et al.* (2003b).

The problem of post-pruning a hierarchy in order to increase its usability is explored in research related to modeling the user interface. Previous work (Botafogo *et al.*, 1992) showed the importance of a balanced hierarchy for the efficient search through hierarchies of menus. Indeed, even though the generally accepted guidelines for the menu design favor breadth over depth (Kiger, 1984), the problem with the breadth hierarchy in large-scale systems is that the number of items at each level may be overwhelming. Hence, a depth hierarchy that limits the number of items at each level may be more effective. This is the so-called breadth/depth trade-off.

Moreover, organizing unstructured business data in useful hierarchies has recently received more attention in the industry. Although there are methods for automatic hierarchy generation, a resultant hierarchy has to be manually pruned, in order to ensure its usability. The main criterion is the coherence of the hierarchy, which ensures that the hierarchy is closely tailored to the needs of the intended user.

4.6. CONCLUSION

To be effective, ontologies need to change as rapidly as the parts of the world they describe. To make this a low effort for human users of systems such as digital libraries, automated support for management of ontology changes is crucial.

In this chapter, we have presented the state-of-the-art in ontology evolution, considering each of the individual phases of the evolution process. Furthermore, we have described how changes to the underlying data and changes to usage patterns can be used to evolve an ontology. In these ways we can reduce the burden of manual ontology engineering.

REFERENCES

Botafogo RA, Rivlin E, Shneiderman B. 1992. Structural analysis of hypertexts: identifying hierarchies and useful metrics. *ACM Transactions on Information System* 10(2):142–180.

Cimiano P, Völker J. 2005. Text2Onto—A framework for ontology learning and data-driven change discovery. In *Proceedings of the 10th International Conference on Applications of Natural Language to Information Systems (NLDB 2005)*, Vol. 3513 of LNCS. Springer.

Davies J, Studer R, Sure Y, Warren P. 2005. Next Generation Knowledge Management. *BT Technology Journal*, 23(3):175–190.

Ehrig M, Haase P, Hefke M, Stojanovic N. 2005. Similarity for ontologies – a comprehensive framework. In *Proceedings of the 13th European Conference on Information Systems (ECIS2005)*.

Franconi E, Grandi F, Mandreoli F. 2000. A semantic approach for schema evolution and versioning in object-oriented databases. In *Proceedings of the First International Conference on Computational Logic*, Springer, pp 1048–1062.

Haase P, Hotho A, Schmidt-Thieme L, Sure Y. 2005a. Collaborative and usage-driven evolution of personal ontologies. *Proceedings of the Second European Semantic Web Conference (ESWC 2005)*, Vol. 3532 of LNCS, Springer, pp 486–499.

Haase P, van Harmelen F, Huang Z, Stuckenschmidt H, Sure Y. 2005b. A framework for handling inconsistency in changing ontologies. In *Proceedings of the Fourth International Semantic Web Conference (ISWC2005)*, Vol. 3729 of LNCS, Springer, pp 353–367.

Haase P, Stojanovic L. 2005. Consistent Evolution of OWL Ontologies. In *Proceedings of the Second European Semantic Web Conference, Heraklion, Greece, 2005*, Vol. 3532 of LNCS, Springer, pp 182–197.

Kiger JI. 1984. The depth/breadth trade-off in the design of menu-driven user interfaces. *International Journal of Man-Machine Studies* Vol. 20(2):201–213.

Klein M. 2004. Change Management for Distributed Ontologies, PhD thesis, Vrije Universiteit Amsterdam.

Klein M, Noy N. 2003. A Component-Based Framework for Ontology Evolution. In *Proceedings of the IJCAI '03 Workshop on Ontologies and Distributed Systems*.

Maedche A, Motik B, Stojanovic L. 2003. Managing multiple and distributed Ontologies in the Semantic Web. *VLDB Journal* 12(4):286–300.

Mädche A, Volz R. 2001. The ontology extraction and maintenance framework text-to-onto. *Proceedings of the ICDM'01 Workshop on Integrating Data Mining and Knowledge Management*.

Ognyanov D, Kiryakov A. 2002. Tracking changes in RDF(S) repositories. *Proceedings of the 13th International Conference on Knowledge Engineering and Knowledge Management. Ontologies and the Semantic Web (EKAW 2002)*, Vol. 2473 of LNCS/LNAI, Springer, pp 373–378.

Pons A, Keller R. 1997. Schema evolution in object databases by catalogs. *Proceedings of the International Database Engineering and Applications Symposium (IDEAS'97)*, pp 368–376.

Staab S, Studer R (Eds). 2004. *Handbook on Ontologies*. Springer: Heidelberg.

Stojanovic L. 2004. Methods and Tools for Ontology Evolution. PhD thesis, University of Karlsruhe.

Stojanovic L, Mädche A, Motik B, Stojanovic N. 2002. User-driven ontology evolution management. In *Proceedings of the European Conference of Knowledge Engineering and Management (EKAW 2002)*, Vol. 2473 of LNCS/LNAI, Springer.

Stojanovic L, Maedche A, Stojanovic N, Studer R. 2003a. Ontology evolution as reconfiguration-design problem solving. In *Proceedings of KCAP 2003*, ACM, pp 162–171.

Stojanovic L, Stojanovic N, Gonzalez J, Studer R. 2003b. OntoManager—A System for the usage-based Ontology Management. In *Proceedings of the CoopIS/DOA/ ODBASE 2003 Conference*, Vol. 2888 of LNCS, Springer, pp 858–875.

Sure Y, Erdmann M, Angele J, Staab S, Studer R, Wenke D. 2002a. OntoEdit: Collaborative ontology Engineering for the Semantic Web. In *Proceedings of the First International Semantic Web Conference 2002 (ISWC 2002)*, Vol. 2342 of LNCS, Springer, pp 221–235.

Sure Y, Staab S, Studer R. 2002b. Methodology for development and employment of ontology based knowledge management applications. *SIGMOD Record*, 31(4):18–23.

Sure Y, Studer R. 2005. Semantic web technologies for digital libraries. *Library Management* 26(4/5):190–195.

Tempich C, Pinto HS, Sure Y, Staab S. 2005. An Argumentation Ontology for DIstributed, Loosely-controlled and evolvInG Engineering processes of oNTologies (DILIGENT). In *Proceedings of the Second European Semantic Web Conference (ESWC 2005)*, Vol. 3532 of LNCS, Springer, pp 241–256.

5

Reasoning With Inconsistent Ontologies: Framework, Prototype, and Experiment*

Zhisheng Huang, Frank van Harmelen and Annette ten Teije

Classical logical inference engines assume the consistency of the ontologies they reason with. Conclusions drawn from an inconsistent ontology by classical inference may be completely meaningless. An inconsistency reasoner is one which is able to return meaningful answers to queries, given an inconsistent ontology. In this chapter, we propose a general framework for reasoning with inconsistent ontologies. We present the formal definitions of soundness, meaningfulness, local completeness, and maximality of an inconsistency reasoner. We propose and investigate a pre-processing algorithm, discuss the strategies of inconsistency reasoning based on pre-defined selection functions dealing with concept relevance. We have implemented a system called PION (Processing Inconsistent ONtologies) for reasoning with inconsistent ontologies. We discuss how the syntactic relevance can be used for PION. In this chapter, we also report the preliminary experiments with PION.

5.1. INTRODUCTION

The Semantic Web is characterized by scalability, distribution, and joint author-ship. All these characteristics may introduce inconsistencies.

---

*This chapter is an extended and revised version of the paper 'Reasoning with Inconsistent Ontologies' appeared in the Proceedings of the 19th Joint Conference on Artificial Intelligence (IJCAI'05), 2005, pp 454–459.

Semantic Web Technologies: Trends and Research in Ontology-based Systems
John Davies, Rudi Studer, Paul Warren © 2006 John Wiley & Sons, Ltd

Limiting the language expressivity with respect to negation (such as RDF and RDF Schema, which do not include negation) can avoid inconsistencies to a certain extent. However, the expressivity of these languages is too limited for many applications. In particular, OWL is already capable of expressing inconsistencies (McGuinness and van Harmelen, 2004).

There are two main ways to deal with inconsistency. One is to diagnose and repair it when we encounter inconsistencies. Schlobach and Cornet (2003) propose a nonstandard reasoning service for debugging inconsistent terminologies. This is a possible approach, if we are dealing with one ontology and we would like to improve this ontology. Another approach is to simply live with the inconsistency and to apply a nonstandard reasoning method to obtain meaningful answers. In this chapter, we will focus on the latter, which is more suitable for the setting in the web area. For example, in a typical Semantic Web setting, one would be importing ontologies from other sources, making it impossible to repair them. Also the scale of the combined ontologies may be too large to make repair effective.

Logical entailment is the inference relation that specifies which consequences can be drawn from a logical theory. A logical theory is inconsistent if it contains a contradiction: for some specific statement A, both A and its negation *not* A are consequences of the theory. As is well known, the classical entailment in logics is *explosive*: any formula is a logical consequence of a contradiction. Therefore, conclusions drawn from an inconsistent knowledge base by classical inference may be completely meaningless. In this chapter, we propose a general framework for reasoning with inconsistent ontologies. We investigate how a reasoner with inconsistent ontologies can be developed for the Semantic Web. The general task of a reasoner with inconsistent ontologies is: given an inconsistent ontology, the reasoner should return *meaningful* answers to queries. In Section 5.4, we will provide a formal definition about meaningfulness.

This chapter is organized as follows: Section 5.2 discusses existing general approaches to reasoning with inconsistency. Section 5.3 overviews inconsistency in the Semantic Web by examining several typical examples and scenarios. Section 5.4 proposes a general framework of reasoning with inconsistent ontologies. A crucial element of this framework is so-called selection functions. Section 5.5 examines selection functions which are based on concept relevance. Section 5.6 presents the strategies and algorithms for processing inconsistent ontologies. Section 5.7 investigates how a selection function can be developed by a syntactic relevance relation. Section 5.8 describes a prototype of PION and report the experiments with PION. Section 5.9 discusses further work and concludes the chapter.

5.2. BRIEF SURVEY OF APPROACHES TO REASONING WITH INCONSISTENCY

5.2.1. Paraconsistent Logics

Reasoning with inconsistency is a well-known topic in logics and AI. Many approaches have been proposed to deal with inconsistency (Benferhat and Garcia, 2002; Beziau, 2000; Lang and Marquis, 2001). The development of paraconsistent logics was initiated to challenge the 'explosive' problem of the standard logics. Paraconsistent logics (Beziau, 2000) allow theories that are inconsistent but nontrivial. There are many different paraconsistent logics, each of which weaken traditional logic in a different way. *Nonadjunctive systems* block the general inference $a, b \models a \wedge b$, so that in particular the combination of a and $\neg a$ no longer entails $a \wedge \neg a$. *Relevace logics* aim to block the explosive inference $a \wedge \neg a \models b$ by requiring that the premises of an entailment must somehow be 'relevant' to the conclusion. In the propositional calculus, this involves requiring that premises and conclusion share atomic sentences, which is not the case in the latter formula.

Many relevant logics are *multi-valued logics*. They are defined on a semantics which allows both a proposition and its negation to hold for an interpretation. Levesque's (1989) limited inference allows the interpretation of a language in which a truth assignment may map both a proposition l and its negation $\neg l$ to true. Extending the idea of Levesque's limited inference, Schaerf and Cadoli (1995) propose S-3-entailment and S-1-entailment for approximate reasoning with tractable results. The main idea of Schaerf and Cadoli's approach is to introduce a subset S of the language, which can be used as a parameter in their framework and allows their reasoning procedure to focus on a part of the theory while the remaining part is ignored. However, how to construct and extend this subset S in specific scenario's is still an open question (the problem of finding a general optimal strategy for S is known to be intractable).

Based on Schaerf and Cadoli's S-3-entailment, Marquis and Porquet (2003) present a framework for reasoning with inconsistency by introducing a family of resource-bounded paraconsistent inference relations. In Marquis and Porquet's approach, consistency is restored by removing variables from the approximation set S instead of removing some explicit beliefs from the belief base, like the standard approaches do in belief revision. Their framework enables some forms of graded paraconsistency by explicit handling of preferences over the approximation set S. Marquis and Porquet (2003) propose several policies, for example, the linear order policy and the lexicographic policy, for the preference handling in paraconsistent reasoning.

5.2.2. Ontology Diagnosis

As mentioned in the introduction, an alternative approach to deal with inconsistencies is to repair them before reasoning, instead of reasoning in the presence of the inconsistencies. A long standing tradition in Artificial Intelligence is that of belief revision, which we will discuss below. A more recent branch of work is explicitly tailored to diagnosis and repair of ontologies in particular. The first in this line was done by Schlobach and Cornet (2003), who aimed at identifying a minimal subset of Description Logic axioms that is responsible for an inconsistency (i.e., such a minimal subset is inconsistent, but removal of any single axiom from the set makes the inconsistency go away). In later works by Friedrich and Shchekotykhin (2005) and Schlobach (2005b), this approach has been extended to deal with richer Description Logics, and has been rephrased in terms of Reiter's (1987) general theory of model-based diagnosis.

5.2.3. Belief Revision

Belief revision is the process of changing beliefs to take into account a new piece of information.

What makes belief revision nontrivial is that several different ways for performing this operation may be possible. For example, if the current knowledge includes the three facts a, b, and $a \wedge b \rightarrow c$, the introduction of the new information $\neg c$ can be done preserving consistency only by removing at least one of the three facts. In this case, there are at least three different ways for performing revision. In general, there may be several different ways for changing knowledge.

The main assumption of belief revision is that of minimal change: the knowledge before and after the change should be as similar as possible. The AGM postulates (Alchourron *et al.*, 1985)[1] are properties that an operator that performs revision should satisfy in order for being considered rational. Revision operators that satisfy the AGM postulates are computationally highly intractable. In an attempt to avoid this, Chopra *et al.* (2000) incorporate the local change of belief revision and relevance sensitivity by means of Schaerf and Cadoli's approximate reasoning method, and show how relevance can be introduced for approximate reasoning in belief revision. Incidently, recent work by Flouris *et al.* (2005) has shown that the AGM theory in its original form is not applicable to restricted logics such as the Description Logics that underly OWL, and that it is not trivial to find alternative formulations of the AGM postulates that would work for OWL.

---

[1]Named after the names of their proponents, Alchourron, Gardenfors, and Makinson.

5.2.4. Synthesis

Various approaches discussed above (Marquies' paraconsistent logic and Chopra's local belief revision) depending on syntactic selection procedures for extending the approximation set. Our approach borrows some ideas from Schaerf and Cadoli's approximation approach, Marquis and Porquet's paraconsistent reasoning approach, and Chopra, Parikh, and Wassermann's relevance approach. However, our main idea is relatively *simple*: given a selection function, which can be defined on the syntactic or semantic relevance, like those have been used in computational linguistics, we select some consistent subtheory from an inconsistent ontology. Then we apply standard reasoning on the selected subtheory to find meaningful answers. If a satisfying answer cannot be found, the relevance degree of the selection function is made less restrictive (see later sections for precise definitions of these notions) thereby extending the consistent subtheory for further reasoning.

5.3. BRIEF SURVEY OF CAUSES FOR INCONSISTENCY IN THE SEMANTIC WEB

In the Semantic Web, inconsistencies may easily occur, sometimes even in small ontologies. Here are several scenarios which may cause inconsistencies:

5.3.1. Inconsistency by Mis-representation of Default

When a knowledge engineer specifies an ontology statement, she/he has to check carefully that the new statement is consistent, not only with respect to existing statements, but also with respect to statements that may be added in the future, which of course may not always be known at that moment. This makes it very difficult to maintain consistency in ontology specifications. Just consider a situation in which a knowledge engineer wants to create an ontology about animals:[2]

$Bird \sqsubseteq Animal$ (Birds are animals),
$Bird \sqsubseteq Fly$ (Birds are flying animals).

Although the knowledge engineer may realize that 'birds can fly' is not generally valid, he still wants to add it if he does not find any counterexample in the current knowledge base because flying is one of

---

[2]Since we are dealing with (simple) ontological examples, we will adopt the notation from Description Logic, underlying the OWL language.

the main features of birds. An ontology about birds without talking about flying is not satisfactory.

Later on, one may want to extend the ontology with the following statements:

$Eagle \sqsubseteq Bird$ (Eagles are birds),
$Penguin \sqsubseteq Bird$ (Penguins are birds),
$Penguin \sqsubseteq \neg Fly$ (Penguins are not flying animals).

The concept $Penguin$ in that ontology of birds is already unsatisfiable because it implies penguins can both fly and not fly. This would lead to an inconsistent ontology when there exists an instance of the concept $Penguin$. One may remove the axiom 'birds can fly' from the existing ontology to restore consistency. However, this approach is not reliable because of the following reasons: (a) it is hard to check that the removal would not cause any significant information loss in the current ontology, (b) one may not have the authority to remove statements which have been created in the current knowledge base, (c) it may be difficult to know which part of the existing ontology can be removed if the knowledge base is very large. One would not blame the knowledge engineer for the creation of the axiom 'birds are flying animals' at the beginning without considering future extensions because it is hard for the knowledge engineer to do so.

One may argue that the current ontology languages and their counterparts in the Semantic Web cannot be used to handle this kind of problems because it requires nonmonotonic reasoning. The statement $Birds\ can\ fly$ has to be specified as a default. The ontology language OWL cannot deal with defaults. We have to wait for an extension of OWL to accommodate nonmonotonic logic. It is painful that we cannot talk about birds (that can fly) and penguins (that cannot fly) in the same ontology specification. An alternative approach is to divide the inconsistent ontology specification into multiple ontologies or modular ontologies to maintain their local consistency, like one that states 'birds can fly,' but does not talk about penguins, and another one that specifies penguins, but never mentions that 'birds can fly.' However, the problem for this approach is still the same as other ones. Again, an ontology about birds that cannot talk about both 'birds can fly' and penguins is not satisfactory.

Another typical example is the MadCows ontolog[3] in which MadCow is specified as a Cow which eats brains of sheep, whereas a Cow is considered as a vegetarian by default as follows:

$Cow \sqsubseteq Vegetarian$ (Cows are vegetarians),
$MadCow \sqsubseteq Cow$ (MadCows are cows),

---

[3]http://www.daml.org/ontologies/399

$MadCow \sqsubseteq \exists Eat.(Brain \sqcap \exists Partof.\ Sheep)$ (MadCows eat brains of sheep, parts of animals).

$Sheep \sqsubseteq Animals$ (Sheep are animals),

$Vegetarians \sqsubseteq \forall Eat.(\neg Animal \sqcup \neg (\exists Partof.\ Animal))$ Vegetarians never eat animals or parts of animals).

In order to make the MadCow Ontology consistent, we have to remove at least one of the statements above from the ontology. Namely, either we would not claim that 'Cows are vegetarian,' or claims that 'MadCows are cow,' or claims that 'Sheep are animals,' or claims that 'Vegetarians never eat animals or parts of animals.' It is quite difficult to decide which statement should be removed, for all are important for a sufficient specification of a MadCow Ontology. We expect a reasoner with inconsistent ontologies can do better work to avoid reparing this ontology.

5.3.2. Inconsistency Caused by Polysemy

Polysemy refers to the concept of words with multiple meanings. One should have a clear understanding of all the concepts when an ontology is formally specified. Here is an example of an inconsistent ontology which is caused by polysemy:

$MarriedWoman \sqsubseteq Woman$ (A married woman is a woman),

$MarriedWoman \sqsubseteq \neg Divorcee$ (A married woman is not a divorcee),

$Divorcee \sqsubseteq HadHusband \sqcap \neg HasHusband$ (A divorcee had a husband and has no husband),

$HasHusband \sqsubseteq MarriedWoman$ (HasHusband means married),

$HadHusband \sqsubseteq MarriedWoman$ (HadHusband means married).

In the ontology specification above, the concepts 'Divorcee' is unsatisfiable because of the misuse of the word 'MarriedWoman.' Therefore, one has to carefully check if there is some misunderstanding with respect to concepts that have been used in the ontology, which may become rather difficult when an ontology is large.

5.3.3. Inconsistency through Migration from Another Formalism

When an ontology specification is migrated from other data sources, inconsistencies may occur. As it has been found by Schlobach and

Cornet (2003), the high number of unsatisfiable concepts in the Description Logic terminology for DICE is due to the fact that it has been created by migration from a frame-based terminological system.[4] In order to make the semantics as explicit as possible, a very restrictive translation has been chosen to highlight as many ambiguities as possible. Schlobach and Cornet (2003) show the following inconsistent ontology specification:

$Brain \sqsubseteq CentralNervousSystem$	(A brain is a central nervous system),
$Brain \sqsubseteq BodyPart$	(A brain is a body part),
$CentralNervousSystem \sqsubseteq NervousSystem$	(A central nervous system is a nervous system),
$BodyPart \sqsubseteq \neg NervousSystem$	(A body part is not a nervous system).

This ontology is inconsistent because a *Brain* is both a *BodyPart* and a *CentralNervousSystem*, and therefore also a *NervousSystem*, but this is inconsist with the axiom that *BodyPart* and *NervousSystem* are disjoint.

5.3.4. Inconsistency Caused by Multiple Sources

When a large ontology specification is generated from multiple sources, in paricular when these sources are created by several authors, inconsistencies easily occur.

Hameed *et al.* (2003) propose approaches of ontology reconciliation, and discuss how consistency should be maintained and how inconsistency may be created from multiple sources. According to Hameed *et al.* (2003), there are three possibilities for ontology reconciliation: merging, aligning, or integrating. No matter whether a new ontology is generated by merging or integrating multiple sources, in both cases general consistency objectives are rather difficult to achieve. Note that the above-mentioned categories (mis-representation of defaults, polysemy, migration, multiple sources) do not exclude each other. When we examine an inconsistent ontology which is generated from multiple sources, we may find that it contains several cases of polysemy, or some other inconsistency. The list above is also not exhaustive. There are many other cases that may cause the inconsistency, like inconsistency caused by ambiguities, inconsistency caused by lacking global checking, etc. We do not discuss a complete list in this chapter, but we have just aimed to show the urgency of the problem of reasoning with inconsistent ontologies.

---

[4]DICE stands for 'Diagnoses for Intensive Care Evaluation.'

5.4. REASONING WITH INCONSISTENT ONTOLOGIES

In this section, we present our framework for reasoning with inconsistent ontologies. First we will introduce some definitions and terminology for inconsistency reasoners, for instance what do we mean with a 'meaningful' answers to a query. We continue with identifying when to use the reasoner for inconsistent ontologies.

5.4.1. Inconsistency Detection

We do not restrict ontology specifications to a particular language (although OWL and its underlying description logic are the languages we have in mind). In general, an ontology language can be considered to be a set of formulas that is generated by a set of syntactic rules. Thus, we can consider an ontology specification as a formula set. We use a nonclassical entailment for inconsistency reasoning. In the following, we use $\models$ to denote the classical entailment, and use $\approx$ to denote some nonstandard inference relation, which may be parameterized to remove ambiguities.

With classical reasoning, a query ϕ given an ontology Σ can be expressed as an evaluation of the consequence relation $\Sigma \models \phi$. There are only two answers to that query: either 'yes' ($\Sigma \models \phi$), or 'no' ($\Sigma \models/= \phi$). A 'yes' answer means that ϕ is a logical consequence of Σ. A 'no' answer, however, means that ϕ cannot be deduced from Σ because we usually do not adopt the closed world assumption when using an ontology. Hence, a 'no' answer does not imply that the negation of ϕ holds given an ontology Σ. For reasoning with inconsistent ontologies, it is more suitable to use Belnap's (1977) four-valued logic to distinguish the following four epistemic states of the answers:

Definition 1.

(a) *Over-determined:* $\Sigma \approx \phi$ *and* $\Sigma \approx \neg\phi$.
(b) *Accepted:* $\Sigma \approx \phi$ and $\Sigma \not\approx \neg\phi$.
(c) *Rejected:* $\Sigma \not\approx \phi$ and $\Sigma \approx \neg\phi$.
(d) *Undetermined:* $\Sigma \not\approx \phi$ and $\Sigma \not\approx \neg\phi$.

To make sure reasoning is reliable when it is unclear if an ontology is consistent or not, we can use the decision tree that is depicted in Figure 5.1. For a query ϕ we test both the consequences $\Sigma \models \phi$ and $\Sigma \models \neg\phi$ using classical reasoning. In case different answers are obtained, that is both 'yes' and 'no,' the ontology Σ must be consistent and the answer to $\Sigma \models \phi$ can be returned. In case of two answers that are the same the ontology is either incomplete, that is when both answers are 'no,' or the ontology is inconsistent, that is when both answers are 'yes.' When the ontology

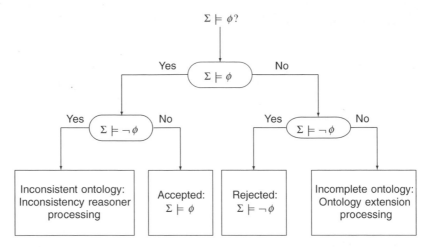

Figure 5.1 Decision tree for obtaining reliable reasoning with an inconsistent ontology.

turns out to be incomplete, either an 'undetermined' answer can be returned or additional information I can be gathered to answer the query $\Sigma \cup I \models \phi$, but this falls outside the scope of the chapter. When the ontology turns out to be inconsistent some inconsistency reasoner can be called upon to answer the query $\Sigma \approx \phi$.

5.4.2. Formal Definitions

For an inconsistency reasoner it is expected that it is able to return meaningful answers to queries, given an inconsistent ontology. In the case of a consistent ontology Σ, classical reasoning is sound, that is a formula ϕ deduced from Σ holds in every model of Σ. This definition is not preferable for an inconsistent ontology Σ as every formula follows from it using classical entailment. However, often only a small part of Σ has been incorrectly constructed or modeled, while the remainder of Σ is correct. Therefore, we propose the following definition of soundness.

Definition 2 (Soundness). *An inconsistency reasoner $\approx$ is sound if the formulas that follow from an inconsistent theory Σ follow from a consistent subtheory of Σ using classical reasoning, namely, the following condition holds:*

$$\Sigma \approx \phi \Rightarrow (\exists \Sigma' \subseteq \Sigma)\ (\Sigma' \not\models \bot \text{ and } \Sigma' \models \phi).$$

In other words, the $\approx$ consequences must be justifiable on the basis of a consistent subset of the theory. Note however, that in the previous definition the implication should *not hold* in the opposite direction. If the implication would also hold in the opposite direction it would lead to an inconsistency reasoner, which returns inconsistent answers. For example if $\{a, \neg a\} \subseteq \Sigma$, then the inconsistency reasoner would return that both a and $\neg a$ hold given Σ, which is something we would like to prevent. Hence, the inconsistency reasoner should not return answers that follow from *any* consistent subset of Σ, but from *specifically chosen* subsets of Σ. In other words, the selection of specific subsets on which $\approx$ will be based is an integral part of the definition of an inconsistency reasoner, and will be discussed in more detail in the next section.

Definition 3 (Meaningfulness). *An answer given by an inconsistency reasoner is meaningful iff it is consistent and sound. Namely, it requires not only the soundness condition, but also the following condition:*

$$\Sigma \approx \phi \Rightarrow \Sigma \not\approx \neg\phi.$$

An inconsistency reasoner is said to be meaningful iff all of the answers are meaningful.

Because of inconsistencies, classical completeness is impossible. We suggest the notion of local completeness:

Definition 4 (Local Completeness). *An inconsistency reasoner is locally complete with respect to a consistent subtheory Σ' iff for any formula ϕ, the following condition holds:*

$$\Sigma' \models \phi \Rightarrow \Sigma \approx \phi.$$

Since the condition can be represented as:

$$\Sigma \not\approx \phi \Rightarrow \Sigma' \not\models \phi,$$

local completeness can be considered as a complement to the soundness property.

An answer to a query ϕ on Σ is said to be locally complete with respect to a consistent set Σ' iff the following condition holds:

$$\Sigma' \models \phi \Rightarrow \Sigma \approx \phi.$$

Definition 5 (Maximality). *An inconsistency reasoner is maximally sound iff for any theory Σ there is a maximal consistent subtheory Σ' such that the classical consequence set of Σ' is the same as the consequence set of the inconsistency reasoner on the full Σ:*

$$\exists(\Sigma' \subseteq \Sigma)((\Sigma' \not\models \bot) \wedge (\forall\Sigma'' \supset \Sigma' \wedge \Sigma'' \subseteq \Sigma)(\Sigma'' \models \bot) \wedge \forall\phi(\Sigma' \models \phi \Leftrightarrow \Sigma \approx \phi)).$$

In other words, the inconsistency reasoner computes precisely the classical consequences of a maximal consistent subtheory.

We use the same condition to define the maximality for an answer, like we do for local completeness.

Definition 6 (Local Soundness). *An answer to a query* $\Sigma \mathrel{\vdash\mkern-10mu\approx} \phi$ *is said to be locally sound with respect to a consistent set* $\Sigma' \subseteq \Sigma$, *iff the following condition holds:*

$$\Sigma \mathrel{\vdash\mkern-10mu\approx} \phi \Rightarrow \Sigma' \models \phi.$$

Namely, for any positive answer, it should be implied by the given consistent subtheory Σ' under the standard entailment.

Proposition 1.

(a) *Local Soundness implies Soundness and Meaningfulness.*
(b) *Maximality implies Local Completeness.*

Given a query, there might exist more than one maximal consistent subset and more than one locally complete consistent subset. Such different maximally consistent subsets may give different $\mathrel{\vdash\mkern-10mu\approx}$ consequences for a given query ϕ. Therefore, arbitrary (maximal) consistent subsets may not be very useful for the evaluation of a query by some inconsistency reasoner. the consistent subsets should be chosen on structural or semantic grounds indicating the relevance of the chosen subset with respect to some query.

5.5. SELECTION FUNCTIONS

An inconsistency reasoner uses a selection function to determine which consistent subsets of an inconsistent ontology should be considered in its reasoning process. The general framework is independent of the particular choice of selection function. The selection function can either be based on a syntactic approach, like Chopra *et al.* (2000) syntactic relevance, or based on semantic relevance like for example in computational linguistics as in Wordnet (Budanitsky and Hirst, 2001).

Given an ontology (i.e., a formula set) Σ and a query ϕ, a selection function s is one which returns a subset of Σ at the step $k > 0$. Let L be the ontology language, which is denoted as a formula set. We have the general definition about selection functions as follows:

Definition 7 (Selection Functions). *A selection function s is a mapping*

$s : \mathcal{P}(L) \times L \times \mathbb{N} \rightarrow \mathcal{P}(L)$ *such that* $s(\Sigma, \phi, k) \subseteq \Sigma$.

In this chapter we will consider only monotonic selection functions.

Definition 8. *A selection function s is called* monotonic *if the subsets it selects monotonically increase or decrease, that is,* $s(\Sigma, \phi, k) \subseteq s(\Sigma, \phi, k + 1)$, *or vice versa.*

For monotonically increasing selection functions, the initial set is either an empty set, that is, $s(\Sigma, \phi, 0) = \emptyset$, or a fixed set Σ_0 when the locality is required. For monotonically decreasing selection functions, usually the initial set $s(\Sigma, \phi, 0) = \Sigma$. The decreasing selection functions will reduce some formulas from the inconsistent set step by step until they find a maximally consistent set.

Monotonically increasing selection functions have the advantage that they do not have to return *all* subsets for consideration at the same time. If a query $\Sigma \mathrel{\vDash\mkern-14mu\raise0.3ex\hbox{$\approx$}} \phi$ can be answered after considering some consistent subset of the ontology Σ for some value of k, then other subsets (for higher values of k) do not have to be considered any more because they will not change the answer of the inconsistency reasoner.

5.6. STRATEGIES FOR SELECTION FUNCTIONS

A reasoner that uses a monotonically increasing/decreasing selection function will be called a reasoner that uses a *linear extension strategy* and a *linear reduction strategy*, respectively.

A linear extension strategy is carried out as shown in Figur 5.2. Given a query $\Sigma \mathrel{\vDash\mkern-14mu\raise0.3ex\hbox{$\approx$}} \phi$, to the initial consistent subset Σ' is set. Then the selection function is called to return a consistent subset Σ'', which Extends Σ', that is, $\Sigma' \subset \Sigma'' \subseteq \Sigma$ for the linear extension strategy. If the selection function

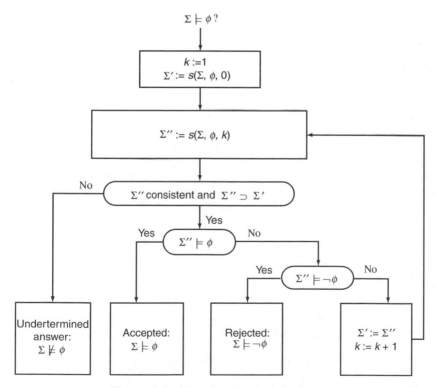

Figure 5.2 Linear extension strategy.

cannot find a consistent superset of Σ', the reasoner returns the answer 'undetermined' (i.e., unknown) to the query. If the set Σ'' exists, a classical reasoner is used to check if $\Sigma'' \models \phi$ holds. If the answer is 'yes,' the reasoner returns the 'accepted' answer $\Sigma \approx \phi$. If the answer is 'no,' the reasoner further checks the negation of the query $\Sigma'' \models \neg\phi$. If the answer is 'yes,' the reasoner returns the 'rejected' answer $\Sigma \not\approx \neg\phi$, otherwise the current result is undetermined (Definition 1), and the whole process is repeated by calling the selection function for the next consistent subset of Σ which extends Σ''.

It is clear that the linear extension strategy may result in too many 'undetermined' answers to queries when the selection function picks the wrong sequence of monotonically increasing subsets. it would therefore be useful to measure the succesfulness of (linear) extension strategies. Notice that this depends on the choice of the monotonic selection function.

In general, one should use an extension strategy that is not over-determined and not undetermined. For the linear extension strategy, we can prove that the following properties hold:

Proposition 2 (Linear Extension). *An inconsistency reasoner using a linear extension strategy satisfies the following properties:*

(a) *never over-determined,*
(b) *may be undetermined,*
(c) *always sound,*
(d) *always meaningful,*
(e) *always locally complete,*
(f) *may not be maximal,*
(g) *always locally sound.*

Therefore, a reasoner using a linear extension strategy is useful to create meaningful and sound answers to queries. It is always locally sound and locally complete with respect to a consistent set Σ_0 if the selection function always starts with the consistent set Σ_0 (i.e., $s(\Sigma, \phi, 0) = \Sigma_0$). Unfortunately it may not be maximal.

We call this strategy a *linear one* because the selection function only follows one possible 'extension chain' for creating consistent subsets. The advantages of the linear strategy is that the reasoner can always focus on the current working set Σ'. The reasoner does not need to keep track of the extension chain. The disadvantage of the linear strategy is that it may lead to an inconsistency reasoner that is undetermined. There exists other strategies which can improve the linear extension approach, for example, by backtracking and heuristics evaluation. We are going to discuss a backtracking strategy in Section 5.7. The second reason why we call the strategy linear is that the computational complexity of the strategy is linear with respect to the complexity of the ontology reasoning (Huang *et al.*, 2005).

5.7 SYNTACTIC RELEVANCE-BASED SELECTION FUNCTIONS

As we have pointed out in Section 5.4, the definition of the selection function should be independent of the general procedure of the inconsistency processing (i.e., strategy). Further research will focus on a formal development of selection functions. However, we would like to point out that there exist several alternatives which can be used for an inconsistency reasoner.

Chopra *et al.* (2000) propose syntactic relevance to measure the relationship between two formulas in belief sets, so that the relevance can be used to guide the belief revision based on Schaerf and Cadoli's method of approximate reasoning. We will exploit their relevance measure as selection function and illustrate them on two examples.

Definition 9 (Direct Relevance and k-Relevance (Chopra *et al.* 2000)). *Given a formula set Σ, two atoms p, q are directly relevant, denoted by $R(p, q, \Sigma)$ if there is a formula $\alpha \in \Sigma$ such that p, q appear in α. A pair of atoms p and q are k-relevant with respect to Σ if there exist $p_1, p_2, \cdots, pk \in \mathcal{L}$ such that:*

- *p, p_1 are directly relevant;*
- *p_i, p_{i+1} are directly relevant, $i = 1, \cdots, k - 1$;*
- *p_k, q are directly relevant.*

The notions of relevance are based on propositional logics. However, ontology languages are usually written in some subset of first order logic. It would not be too difficult to extend the ideas of relevance to those first-order logic-based languages by considering an atomic formula in first-order logic as a primitive proposition in propositional logic.

Given a formula ϕ, we use $I(\phi)$, $C(\phi)$, $R(\phi)$ to denote the sets of individual names, concept names, and relation names that appear in the formula ϕ, respectively.

Definition 10 (Direct Relevance). *Two formula ϕ and ψ are directly relevant if there is a common name which appears both in formula ϕ and formula ψ, that is $I(\phi) \cap I(\psi) \neq \varnothing \vee C(\phi) \cap C(\psi) \neq \varnothing \vee R(\phi) \cap R(\psi) \neq \varnothing$.*

Definition 11 (Direct Relevance to a Set). *A formula ϕ is relevant to a set of formula Σ if there exists a formula $\psi \in \Sigma$ such that ϕ and ψ are directly relevant.*

We can similarly specialize the notion of k-relevance.

Definition 12 (k-Relevance). *Two formulas ϕ, ϕ' are k-relevant with respect to a formula set Σ if there exist formulas $\psi_0, \cdots \psi_k \in \Sigma$ such that ϕ and ψ_0, ψ_0 and ψ_1, $\cdots$, and ψk and ϕ' are directly relevant.*

Definition 13 (k-Relevance to a set). *A formula ϕ is k-relevant to a formula set Σ if there exists formula $\psi \in \Sigma$ such that ϕ and ψ are k-relevant with respect to Σ.*

In inconsistency reasoning we can use syntactic relevance to define a selection function s to extend the query '$\Sigma \approx \phi$?' as follows: We start with the query formula ϕ as a starting point for the selection based on syntactic relevance. Namely, we define:

$$s(\Sigma, \phi, 0) = \emptyset.$$

Then the selection function selects the formulas $\psi \in \Sigma$ which are directly relevant to ϕ as a working set (i.e., $k = 1$) to see whether or not they are sufficient to give an answer to the query. Namely, we define:

$$s(\Sigma, \phi, 1) = \{\psi \in \Sigma \mid \phi \text{ and } \psi \text{ are directly relevant}\}.$$

If the reasoning process can obtain an answer to the query, it stops. otherwise the selection function increases the relevance degree by 1, thereby adding more formulas that are relevant to the current working set. Namely, we have:

$$s(\Sigma, \phi, k) = \{\psi \in \Sigma \mid \psi \text{ is directly relevant to } s(\Sigma, \phi, k-1)\},$$

for $k > 1$. This leads to a 'fan out' behavior of the selection function: the first selection is the set of all formulae that are directly relevant to the query; then all formulae are selected that are directly relevant to that set, etc. This intuition is formalized in the following:

Proposition 3. *The syntactic relevance-based selection function s is monotonically increasing*.

Proposition 4. *If $k \geq 1$, then*

$$s(\Sigma, \phi, k) = \{\phi \mid \phi \text{ is } (k\text{-}1)\text{-relevant to } \Sigma\}$$

The syntactic relevance-based selection functions defined above usually grows up to an inconsistent set rapidly. That may lead to too many undetermined answers. In order to improve it, we require that the selection function returns a consistent subset Σ'' at the step k when $s(\Sigma, \phi, k)$ is inconsistent such that $s(\Sigma, \phi, k-1) \subset \Sigma'' \subset s(\Sigma, \phi, k)$. It is actually a kind of backtracking strategy which is used to reduce the number of undetermined answers to improve the linear extension strategy. We call the procedure an *over-determined processing* (ODP) of the selection function. Note that the over-determined processing does not need to exhaust the powerset of the set $s(\Sigma, \phi, k) - s(\Sigma, \phi, k-1)$ because of the fact that if a consistent set S cannot prove or disprove a query, then nor can any subset of S. Therefore, one approach of ODP is to return just a maximally consistent subset. Let n be $|\Sigma|$ and k be $n - |S|$, that is the cardinality difference between the ontology Σ and its maximal consistent subset S (note that k is usually very small), and let C be the complexity of the consistency checking. The complexity of the over-determined processing is polynomial to the complexity of the consistency checking (Huang *et al.*, 2005).

Note that ODP introduces a degree of non-determinism: selecting different maximal consistent subsets of $s(\Sigma, \phi, k)$ may yield different answers to the query $\Sigma \mathrel{|\!\approx} \phi$. The simplest example of this is $\Sigma = \{\phi, \neg\phi\}$.

5.8. PROTOTYPE OF PION

5.8.1. Implementation

We are implementing the prototype of PION by using SWI-Prolog.[5] PION implements an inconsistency reasoner based on a linear extension strategy and the syntactic relevance-based selection function as discussed in Sections 5.6 and 5.7. PION is powered by XDIG, an extended DIG Description Logic interface for Prolog (Huang and Visser, 2004). PION supports the TELL requests in DIG data format and in OWL, and the ASK requests in DIG data format. A prototype of PION is available for download at the website: http://wasp.cs.vu.nl/sekt/pion.

The architecture of a PION is designed as an extension of the XDIG framework, and is shown in Figure 5.3. A PION consists of the following components:

- *DIG Server:* The standard XDIG server acts as PION's XDIG server, which deals with requests from other ontology applications. It not only supports standard DIG requests, like 'tell' and 'ask,' but also provides additional reasoning facilities, like the identification of the reasoner or change of the selected selection functions.
- *Main Control Component*: The main control component performs the main processing, like query analysis, query pre-processing, and the extension strategy, by calling the selection function and interacting with the ontology repositories.

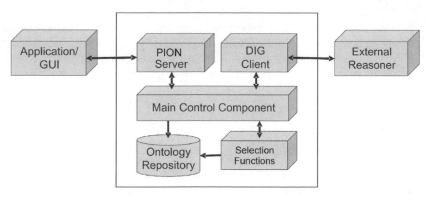

Figure 5.3 Architecture of PION.

---

[5]http://www.swi-prolog.org

- *Selection Functions*: The selection function component is an enhanced component to XDIG, it defines the selection functions that may be used in the reasoning process.
- *DIG Client*: PION's DIG client is the standard DIG client, which calls external description Logic reasoners that support the DIG interface to obtain the standard Description Logic reasoning capabilities.
- *Ontology Repositories*: The ontology repositories are used to store ontology statements, provided by external ontology applications.

The current version of the PION implements a reasoner based on a linear extension strategy and a k-relevance selection function as discussed in Sections 5.2 and 5.5. A screenshot of the PION testbed, is shown in Figure 5.4.

5.8.2. Experiments and Evaluation

We have tested the prototype of PION by applying it on several example ontologies. These example ontologies are the bird example, the brain example, the Married-Woman example, and the MadCow Ontology, which are discussed in Section 5.3. We compare PION's answers with their intuitive answers which is supposed by a human to see to what extend PION can provide intended answers.

For a query, there might exist the following difference between an answer by PION and its intuitive answer.

- *Intended Answer*: PION's answer is the same as the intuitive answer.
- *Counter-Intuitive Answer*: PION's answer is opposite to the intuitive answer. Namely, the intuitive answer is 'accepted' whereas PION's answer is 'rejected,' or vice versa.
- *Cautious Answer*: The intuitive answer is 'accepted' or 'rejected,' but PION's answer is 'undetermined.'
- *Reckless Answer*: PION's answer is 'accepted' or 'rejected' whereas the intuitive answer is 'undetermined.' We call it a reckless answer because under this situation PION returns just one of the possible answers without seeking other possibly opposite answers, which may lead to 'undetermined.'

For each concept C in those ontologies, we create an instance 'the_C' on them. We make both a positive instance query and a negative instance query of the instance 'the_C' for some concepts D in the ontologies, like a query is 'the_C a D?' PION test results are shown in Figure 5.5. Of the four test examples, PION can return at least 85.7 % intended answers. Of the 396 queries, PION returns 24 cautious answers or reckless answers, and 2 counter-intuitive answers. However, we would like to point out that the high rate of the intended answers includes many 'undetermined' answers. One interesting (and we believe realistic) property of the Mad

Figure 5.4 PION testbed.

Example	Queries	IA	CA	RA	CIA	IA Rate(%)	ICR Rate(%)
Bird	50	50	0	0	0	100	100
Brain	42	36	4	2	0	85.7	100
MarriedWoman	50	48	0	2	0	96	100
MadCow	254	236	16	0	2	92.9	99

IA, intended answers; CA, cautious answers; RA, reckless answers; CIA, counter-intuitive answers; IA Rate, intended answers (%); ICR rate, IA+CA+RA (%).

Figure 5.5 PION test results.

Cows ontology is that many concepts which are intuitively disjoint (such as cows and sheep) are not actually declared as being disjoint (keep in mind that OWL has an open world semantics, and does not make the unique name assumption). As a result, many queries such as 'is the_cow a sheep' are indeed undetermined on the basis of the ontology, and PION correctly reports them as undetermined. The average time cost of the tested queries is about 5 seconds even on a low-end PC (with 550 mHz CPU, 256 MB memory under Windows 2000).

The counter-intuitive results occur in the MadCows Example. PION returns the 'accepted' answer to the query 'is the_mad_cow a vegetarian?' This counter-intuitive answer results from the weakness of the syntactic relevance-based selection function because it always prefers a shorter relevance path when a conflict occurs. In the mad cow example, the path 'mad cow – cow – vegetarian' is shorter than the path 'mad cow –.eat brain – eat bodypart – sheep are animals – eat animal – NOT vegetarian.' Therefore, the syntactic relevance-based selection function finds a consistent subtheory by simply ignoring the fact 'sheep are animals.' The problem results from the unbalanced specification between Cow and MadCow, in which Cow is directly specified as a vegetarian whereas there is no direct statement 'a MadCow is not a vegetarian.'

There are several alternative approaches to solve this kind of problems. One is to introduce the locality requirement. Namely, the selection function starts with a certain subtheory which must always be selected. For example, the statement 'sheep are animals' can be considered to be a knowledge statement which cannot be ignored. Another approach is to add a shortcut path, like the path 'mad cow – eat animal – NOT vegetarian' to achieve the relevance balance between the concepts 'vegetarian' and NOT vegetarian,' as shown in the second mad cow example of PION testbed. The latter approach can be achieved automatically by accommodation of the semantic relevance from the user queries. The hypothesis is that both concepts appear in a query more frequently, when they are semantically more relevant. Therefore, from a semantic point of view, we can add a relevance shortcut path between strongly relevant concepts.

5.8.3. Future Experiments

As noted in many surveys of current Semantic Web work, most Semantic Web applications to date (including those included in this volume) use rather lightweight ontologies. These lightweight ontologies are often expressed in RDF Schema, which means that by definition they will not contain any inconsistencies. However, closer inspection by Schlobach (2005a) revealed that such lightweight ontologies contain many implicit assumptions (such as disjointness of siblings in the class hierarchy) that have not been modeled explicitly because of the limitations of the lightweight representation language Schlobach's (2005a) study reveals that after making such implicit disjointness assumptions explicit (a process called *semantic clarification*), many of the ontologies do reveal internal inconsistencies. In future experiments, we intend to determine to which extent it is still possible to locally reason in such semantically clarified inconsistent ontologies using the heuristics described in this chapter.

5.9. DISCUSSION AND CONCLUSIONS

In this chapter, we have presented a framework for reasoning with inconsistent ontologies. We have introduced the formal definitions of the selection functions, and investigated the strategies of inconsistency reasoning processing based on a linear extension strategy.

One of the novelties of our approach is that the selection functions depend on individual queries. Our approach differs from the traditional one in paraconsistent reasoning, nonmonotonic reasoning, and belief revision, in which a pre-defined preference ordering for all of the queries is required. This makes our approach more flexible, and less inefficient to obtain intended results. The selection functions can be viewed as ones creating query-specific preference orderings.

We have implemented and presented a prototype of PION. In this chapter, we have provided the evaluation report of the prototype by applying it to the several inconsistent ontology examples. The tests show that our approach can obtain intuitive results in most cases for reasoning with inconsistent ontologies. Considering the fact that standard reasoners always result in either meaningless answers or incoherence errors for queries on inconsistent ontologies, we can claim that PION can do much better because it can provide a lot of intuitive, thus meaningful answers. This is a surprising result given the simplicity of our selection function.

We are also working on a framework for inconsistent ontology diagnosis and repair by defining a number of new nonstandard reasoning services to explain inconsistencies through pinpointing (Schlobach and Huang, 2005). An informed bottom-up approach to calculate minimally inconsistent sets by the support of an external Description Logic reasoner has been proposed in Schlobach and Huang (2005). That approach has been prototypically implemented as the DION (Debugger of Inconsistent Ontologies). DION uses the relevance relation which has been used in PION as its heuristic information to guide the selecting procedure for finding minimally inconsistent sets. That justifies to some extent that the notion of 'concept relevance' is useful for inconsistent ontology processing.

In future work, we are going to test PION with more large-scale ontology examples. We are also going to investigate different approaches for selection functions (e.g., semantic-relevance based) and different extension strategies as alternatives to the linear extension strategy in combination with different selection functions, and test their performance.

ACKNOWLEDGMENT

We are indebted to Peter Haase for so carefully proofreading this chapter.

REFERENCES

Alchourron C, Gaerdenfors P, Makinson D. 1985. On the logic of theory change: partial meet contraction and revision functions. *The Journal of Symbolic Logic* 50: 510–530.

Belnap N. 1977. A useful four-valued logic. In Modern Uses of Multiple-Valued Logic, Reidel, Dordrecht, pp 8–37.

Benferhat S, Garcia L. 2002. Handling locally stratified inconsistent knowledge bases, *Studio Logica*, 77–104.

Beziau J-Y. 2000. What is paraconsistent logic. In Frontiers of Paraconsistent Logic. Research Studies Press: Baldock, pp 95–111.

Budanitsky A, Hirst G. 2001. Semantic distance in wordnet: An experimental, application-oriented evaluation of five measures. In Workshop on WordNet and Other Lexical Resources, Pittsburgh, PA.

Chopra S, Parikh R, Wassermann R. 2000. Approximate belief revision-preliminninary report. *Journal of IGPL*.

Flouris G, Plexousakis D. Antoniou G. 2005. On applying the agm theory to dls and owl. In International Semantic Web Conference, LNCS, Springer verlag.

Friedrich G, Shchekotykhin K. 2005. A general diagnosis method for ontologies. In International Semantic Web Conference, LNCS, Springer Verlag.

Hameed A, Preece A. Sleeman D. 2003. Ontology reconciliation. In *Handbook on Ontologies in Information Systems*. Springer Verlag, pp 231–250.

Huang Z, van Harmelen F, ten Teije A. 2005. Reasoning with inconsistent ontologies. In Proceedings of the International Joint Conference on Artificial Intelligence - IJCAI'05, pp 454-459.

Huang Z, Visser C. 2004. Extended DIG description logic interface support for PROLOG, Deliverable D3.4.1.2, SEKT.

Lang J, Marquis P. 2001. Removing inconsistencies in assumption-based the-ories through knowledge-gathering actions. *Studio, Logica*, 179–214.

Levesque HJ (1989). A Knowledge-level account of abduction. In Proceedings of IJCAI'89, pp 1061–1067.

Marquis P, Porquet N. 2003. Resource-bounded paraconsistent inference. *Annals of Mathematics and Artificial Intelligence*, 349–384.

McGuinness D, van Harmelen F. 2004. Owl web ontology language, Recommendation, W3C. http://www.w3.org/TR/owl-features/.

Reiter R. 1987. A theory of diagnosis from first principles. *Artificial Intelligence Journal* 32:57–96.

Schaerf M, Cadoli M. 1995. Tractable reasoning via approximation. *Artificial Intelligence*, 249–310.

Schlobach S. 2005a. Debugging and semantic clarification by pinpointing. In Proceedings of the European Semantic Web Symposium, Vol. 3532 of *LNCS*, Springer Verlag, pp 226–240.

Schlobach S. 2005b. Diagnosing terminologies. In Proceedings of the Twentieth National Conference on Artificial Intelligence, AAAI'05, AAAI, pp 670–675.

Schlobach S, Cornet R. 2003. Non-standard reasoning services for the debugging of description logic terminologies. In Proceedings of IJCAI 2003'.

Schlobach S, Huang Z. 2005 Inconsistent ontology diagnosis: Framework and prototype, Project Report D3.6.1, SEKT.

6

Ontology Mediation, Merging, and Aligning

Jos de Bruijn, Marc Ehrig, Cristina Feier, Francisco Martín-Recuerda, François Scharffe and Moritz Weiten

6.1. INTRODUCTION

On the Semantic Web, data is envisioned to be annotated using ontologies. Ontologies convey background information which enriches the description of the data and which makes the context of the information more explicit. Because ontologies are *shared* specifications, the same ontologies can be used for the annotation of multiple data sources, not only Web pages, but also collections of XML documents, relational databases, etc. The use of such shared terminologies enables a certain degree of inter-operation between these data sources. This, however, does not solve the integration problem completely, because it cannot be expected that all individuals and organizations on the Semantic Web will ever agree on using one common terminology or ontology (Visser and Cui, 1998; Uschold, 2000). It can be expected that many different ontologies will appear and, in order to enable inter-operation, differences between these ontologies have to be reconciled. The reconciliation of these differences is called *ontology mediation*.

Ontology mediation enables reuse of data across applications on the Semantic Web and, in general, cooperation between different organizations. In the context of semantic knowledge management, ontology mediation is especially important to enable *sharing* of data between heterogeneous knowledge bases and to allow applications to *reuse* data

Semantic Web Technologies: Trends and Research in Ontology-based Systems
John Davies, Rudi Studer, Paul Warren © 2006 John Wiley & Sons, Ltd

from different knowledge bases. Another important application area for ontology mediation is *Semantic Web Services*. In general, it cannot be assumed that the requester and the provider of a service use the same terminology in their communication and thus mediation is required in order to enable communication between heterogeneous business partners.

We distinguish two principled kinds of ontology mediation: *ontology mapping* and *ontology merging*. With ontology mapping, the correspondences between two ontologies are stored separately from the ontologies and thus are not part of the ontologies themselves. The correspondences can be used for, for example, querying heterogeneous knowledge bases using a common interface or transforming data between different representations. The (semi-)automated discovery of such correspondences is called *ontology alignment*.

When performing *ontology merging*, a new ontology is created which is the union of the source ontologies. The merged ontology captures all the knowledge from the original ontologies. The challenge in ontology merging is to ensure that all correspondences and differences between the ontologies are reflected in the merged ontology.

Summarizing, *ontology mapping* is mostly concerned with the representation of correspondences between ontologies; *ontology alignment* is concerned with the discovery of these correspondences; and *ontology merging* is concerned with creating the union of ontologies, based on correspondences between the ontologies. We provide an overview of the main approaches in ontology merging, ontology mapping, and ontology alignment in Section 6.2.

After the survey we present a practical approach to ontology mediation where we describe a language to specify ontology mappings, an alignment method for semi-automatically discovering mappings, a graphical tool for browsing and creating mappings in a user friendly way, in Section 6.3.

We conclude with a summary in Section 6.4.

6.2. APPROACHES IN ONTOLOGY MEDIATION

In this section we give an overview of some of the major approaches in ontology mediation, particularly focusing on ontology mapping, alignment, and merging.

An important issue in these approaches is the location and specification of the overlap and the mismatches between concepts, relations, and instances in different ontologies. In order to achieve a better understanding of the mismatches which all these approaches are trying to overcome, we give an overview of the mismatches which might occur between different ontologies, based on the work by Klein (2001), in Section 6.2.1.

We survey a number of representative approaches for ontology mapping, ontology alignment, and ontology merging in Sections 6.2.2, 6.2.3, and 6.2.4, respectively. For more elaborate and detailed surveys we refer the reader to References (Kalfoglou and Schorlemmer, 2003; Noy, 2004; Doan and Halevy, 2005; Shvaiko and Euzenat, 2005).

6.2.1. Ontology Mismatches

The two basic types of ontology mismatches are: (1) *Conceptualization mismatches*, which are mismatches of different conceptualizations of the same domain and (2) *Explication mismatches*, which are mismatches in the way a conceptualization is specified.

Conceptualization mismatches fall in two categories. A *scope mismatch* occurs when two classes have some overlap in their extensions (the sets of instances), but the extensions are not exactly the same (e.g., the concepts `Student` and `TaxPayer`). There is a mismatch in the *model coverage and granularity* if there is a difference in (a) the part of the domain that is covered by both ontologies (e.g., the ontologies of university employees and students) or (b) the level of detail with which the model is covered (e.g., one ontology might have one concept `Person` whereas another ontology distinguishes between `YoungPerson`, `MiddleAged-Person`, and `OldPerson`).

Explication mismatches fall in three categories. There is (1) a mismatch in the *style of modeling* if either (a) the *paradigm* used to specify a certain concept (e.g., time) is different (e.g., intervals vs. points in time) or (b) the way the *concept is described* differs (e.g., using subclasses vs. attributes to distinguish groups of instances). There is a (2) *terminological mismatch* when two concepts are equivalent, but they are represented using different names (*synonyms*) or when the same name is used for different concepts (*homonyms*). Finally, an (3) *encoding mismatch* occurs when values in different ontologies are encoded in a different way (e.g., using kilometers vs. miles for a distance measure).

6.2.2. Ontology Mapping

An *ontology mapping* is a (declarative) specification of the semantic overlap between two ontologies; it is the output of the mapping process (see Figure 6.1). The correspondences between different entities of the two ontologies are typically expressed using some axioms formulated in a specific mapping language. The three main phases for any mapping process are: (1) mapping discovery, (2) mapping representation, and (3) mapping exploitation/execution. In this section we survey a number existing approaches for ontology mapping, with a focus on the mapping representation aspect.

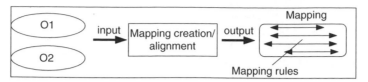

Figure 6.1 Ontology mapping.

A common tendency among ontology mapping approaches is the existence of an ontology of mappings (e.g., MAFRA (Maedche *et al.*, 2002), RDFT (Omelayenko, 2002)), which constitutes the vocabulary for the representation of mappings.

MAFRA (MApping FRAmework for distributed ontologies) (Maedche *et al.*, 2002) supports the interactive, incremental, and dynamic ontology mapping process, where the final purpose of such a process is to support instance transformation. It addresses all the phases of the mapping process: *lift & normalization* (lifting the content of the ontologies to RDF-S and normalization of their vocabularies by eliminating syntactical and lexical differences), *similarity* (computation of the similarities between ontology entities as a support for mapping discovery), *semantic bridging* (establishing correspondences between similar entities, in the form of so-called semantic bridges—defining the mapping), *execution* (exploiting the bridges/mapping for instance transformation), and *post-processing* (revisiting the mapping specification for improvements).

We will focus in the following on the representation of mappings using semantic bridges in MAFRA. The semantic bridges are captured in the Semantic Bridging Ontology (SBO). SBO is a taxonomy of generic bridges; instances of these generic bridges, called *concrete bridges*, constitute the actual concrete mappings. We give an overview of the dimensions along which a bridge can be described in MAFRA, followed by a shallow description of the classes of SBO which allow one to express such bridges.

A bridge can be described along five dimensions:

1. *Entity dimension*: pertains to the entities related by a bridge which may be concepts (modeling classes of objects in the real world), relations, attributes, or extensional patterns (modeling the content of instances).
2. *Cardinality dimension*: pertains to the number of ontology entities at both sides of the semantic bridge (usually 1:n or m:1; m:n is seldom required and it can be usually decomposed into m:1:n).
3. *Structural dimension*: pertains to the way elementary bridges may be combined into a more complex bridge (relations that may hold between bridges: specialization, alternatives, composition, abstraction).
4. *Transformation dimension*: describes how instances are transformed by means of an associated transformation function.

5. *Constraint dimension*: allows one to express conditions upon whose fulfillment the bridge evaluation depends. The transformation rule associated with the bridge is not executed unless these conditions hold.

The abstract class `SemanticBridge` describes a generic bridge, upon which there are no restrictions regarding the entity types that the bridge connects or the cardinality. For supporting *composition*, this class has defined a relation `hasBridge`. The class `SemanticBridgeAlt` supports the *alternative* modeling primitive by grouping several mutually exclusive semantic bridges. The abstract class `SemanticBridge` is further specialized in the SBO according to the entity type: `Relation-Bridge`, `ConceptBridge`, and `AttributeBridge`. `Rule` is a class for describing generic rules. `Condition` and `Transformation` are its subclasses which are responsible for describing the condition necessary for the execution of a bridge and the transformation function of a bridge, respectively. The `Service` class maps the bridge parameters with the transformation procedure arguments to procedures.

RDFT (Omelayenko, 2002) is a mapping meta-ontology for mapping XML DTDs to/and RDF schemas targeted towards business integration tasks. The business integration task in this context is seen as a service integration task, where each enterprise is represented as a Web service specified in WSDL. A conceptual model of WSDL was developed based on RDF Schema extended with the temporal ontology PSL. Service integration is reduced to concept integration; RDFT contains mapping-specific concepts such as events, messages, vocabularies, and XML-specific parts of the conceptual model.

The most important class of the meta-ontology is `Bridge`, which enables one to specify correspondences between one entity and a set of entities or vice versa, depending on the type of the bridge: *one-to-many* or *many-to-one*. The relation between the source and target components of a bridge can be an `EquivalentRelation` (states the equivalence between the two components) or a `VersionRelation` (states that the target set of elements form a later version of the source set of elements, assuming identical domains for the two). This is specified via the bridge property `Relation`. Bridges can be categorized in:

- `RDFBridges`, which are bridges between RDF Schema entities. These can be `Class2Class` or `Property2Property` bridges.
- `XMLBridges`, which are bridges between XML tags of the source/target DTD and the target/source RDF Schema entities. These can be `Tag2-Class`, `Tag2Property`, `Class2Tag`, or `Property2Tag` bridges.
- `Event2Event` bridges, which are bridges that connect two events pertaining to different services. They connect instances of the meta-class `mediator:Event`.

Collections of bridges which serve a common purpose are grouped in a map. When defined in such a way, as a set of bridges, mappings are said to be *declarative*, while *procedural* mappings can be defined by means of an XPath expression for the transformation of instance data.

C-OWL Another perspective on ontology mapping is given by Context OWL (C-OWL) (Bouquet *et al.*, 2004), which is a language that extends the ontology language OWL (Dean and Schreiber, 2004) both syntactically and semantically in order to allow for the representation of *contextual ontologies*. The term contextual ontology refers to the fact that the contents of the ontology are kept local and they can be mapped with the contents of other ontologies via explicit mappings (bridge rules) to allow for a controlled form of global visibility. This is opposed to the OWL importing mechanism where a set of local models is globalized in a unique shared model.

Bridge rules allow connecting entities (concepts, roles, or individuals) from different ontologies that subsume one another, are equivalent, are disjoint or have some overlap. A C-OWL mapping is a set of bridges between two ontologies. A set of OWL ontologies together with mappings between each of them is called a context space.

The local models semantics defined for C-OWL, as opposed to the OWL global semantics, considers that each context uses a local set of models and a local domain of interpretation. Thus, it is possible to have ontologies with contradicting axioms or unsatisfiable ontologies without the entire context space being unsatisfiable.

6.2.3. Ontology Alignment

Ontology alignment is the process of discovering similarities between two source ontologies. The result of a matching operation is a specification of similarities between two ontologies. Ontology alignment is generally described as the application of the so-called *Match* operator (cf. (Rahm and Bernstein, 2001)). The input of the operator is a number of ontology and the output is a specification of the correspondences between the ontologies.

There are many different algorithms which implement the match operator. These algorithms can be generally classified along two dimensions. On the one hand there is the distinction between schema-based and instance-based matching. A schema-based matcher takes different aspects of the concepts and relations in the ontologies and uses some similarity measure to determine correspondence (e.g., (Noy and Musen, 2000b)). An instance-based matcher takes the instances which belong to the concepts in the different ontologies and compares these to discover similarity between the concepts (e.g., (Doan *et al.*, 2004)). On the other hand there is the distinction between element-level and structure-level matching. An element-level matcher compares properties of the particular concept or relation, such as the name, and uses these to find similarities (e.g., (Noy and Musen, 2000b)). A structure-level matcher compares the structure (e.g., the concept hierarchy) of the ontologies to find similarities (e.g., (Noy and Musen, 2000a; Giunchiglia and Shvaiko, 2004)). These matchers can also be combined (e.g., (Ehrig and Staab, 2004; Giunchiglia *et al.*, 2004)). For example, Anchor-PROMPT (Noy and Musen, 2000a), a structure-level matcher, takes as input an initial list of similarities between concepts. The algorithm is then used to find additional similarities, based on the initial similarities and the structure of the ontologies. For a more detailed classification of alignment techniques we refer to Shvaiko and Euzenat (2005). In the following, we give an overview of those approaches.

Anchor-PROMPT (Noy and Musen, 2000a) is an algorithm which aims to augment the results of matching methods which only analyze local context in ontology structures, such as PROMPT (Noy and Musen, 2000b), by finding additional possible points of similarity, based on the structure of the ontologies. The algorithm takes as input two pairs of related terms and analyzes the elements which are included in the path that connects the elements of the same ontology with the elements of the equivalent path of the other ontology. So, we have two paths (one for each ontology) and the terms that comprise these paths. The algorithm then looks for terms along the paths that might be similar to the terms of the other path, which belongs to the other ontology, assuming that the elements of those paths are often similar as well. These new potentially related terms are marked with a similarity score which can be modified during the evaluation of other paths in which these terms occur. Terms with high similar scores will be presented to the user to improve the set of possible suggestions in, for example, a merging process in PROMPT.

GLUE (Doan *et al.*, 2003; 2004) is a system which employs machine-learning technologies to semi-automatically create mappings between heterogeneous ontologies based on instance data, where an ontology is seen as a taxonomy of concepts. GLUE focuses on finding 1-to-1 mappings between concepts in taxonomies, although the authors mention that extending matching to relations and attributes, and involving more complex mappings (such as 1-to-n and n-to-1 mappings) is the subject of ongoing research.

The similarity of two concepts A and B in two taxonomies O1 and O2 is based on the sets of instances that overlap between the two concepts. In order to determine whether an instance of concept B is also an instance of concept A, first a classifier is built using the instances of concept A as the training set. This classifier is now used to classify the instances of concept B. The classifier then decides for each instance of B, whether it is also an instance of A or not.

Based on these classifications, four probabilities are computed, namely, $P(A,B)$, $P(\underline{A},B)$, $P(A,\underline{B})$, and $P(\underline{A},\underline{B})$, where, for example, $P(A,\underline{B})$ is the probability that an instance in the domain belongs to A, but not to B. These four probabilities can now be used to compute the *joint probability distribution* for the concepts A and B, which is a user supplied function with these four probabilities as parameters.

Semantic Matching (Giunchiglia and Shvaiko, 2004) is an approach to matching classification hierarchies. The authors implement a *Match* operator that takes two graph-like structures (e.g., database schemas or ontologies) as input and produces a mapping between elements of the two graphs that semantically correspond to each other.

Giunchiglia and Shvaiko (2004) have argued that almost all earlier approaches to schema and ontology matching have been *syntactic* matching approaches, as opposed to *semantic* matching. In syntactic matching, the labels and sometimes the syntactical structure of the graph are matched and typically some similarity coefficient [0,1] is obtained, which indicates the similarity between the two nodes. Semantic Matching computes a set-based relation between the nodes, taking into account the meaning of each node; the semantics of a node is determined by the label of that node and the semantics of all the nodes which are higher in the hierarchy. The possible relations returned by the Semantic Matching algorithm are *equality* ($=$), *overlap* ($\cap$), *mismatch* ($\perp$), *more general* ($\subseteq$), or *more specific* ($\supseteq$). The correspondence of the symbols with set theory is not a coincidence, since each concept in the classification hierarchies represents a set of documents.

Quick Ontology Mapping (*QOM*) (Ehrig and Staab, 2004; Ehrig and Sure, 2004) was designed to provide an efficient matching tool for on-the-fly creation of mappings between ontologies.

In order to speed up the identification of similarities between two ontologies, QOM does not compare all entities of the first ontology with all entities of the second ontology, but uses heuristics (e.g., similar labels) to lower the number of candidate mappings, that is the number of mappings to compare. The actual similarity computation is done by using a wide range of similarity functions, such as string similarity.

Several of such similarity measures are computed, which are all input to the similarity aggregation function, which combines the individual similarity measures. QOM applies a so-called sigmoid function, which emphasizes high individual similarities and de-emphasizes low individual similarities. The actual correspondences between the entities in the ontologies are extracted by applying a threshold to the aggregated similarity measure. The output of one iteration can be used as part of the input in a subsequent iteration of QOM in order to refine the result. After a number of iterations, the actual table of correspondences between the ontologies is obtained.

6.2.4. Ontology Merging

Ontology merging is the creation of one ontology from two or more source ontologies. The new ontology will unify and in general replace the original source ontologies. We distinguish two distinct approaches in ontology merging. In the first approach the input of the merging

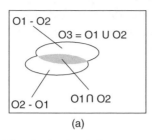
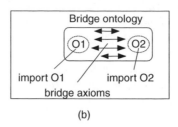

(a)   (b)

Figure 6.2 Output of the merging process. (a) Complete merge and (b) bridge ontology.

process is a collection of ontologies and the outcome is one new, merged, ontology which captures the original ontologies (see Figure 6.2(a)). A prominent example of this approach is PROMPT (Noy and Musen, 2000b), which is an algorithm and a tool for interactively merging ontologies. In the second approach the original ontologies are not replaced, but rather a 'view,' called *bridge ontology*, is created which imports the original ontologies and specifies the correspondences using *bridge axioms*. OntoMerge (Dou *et al.*, 2002) is a prominent example of this approach. OntoMerge facilitates the creation of a 'bridge' ontology which imports the original ontologies and relates the concepts in these ontologies using a number of bridge axioms. We describe the PROMPT and OntoMerge approaches in more detail below.

PROMPT (Noy and Musen, 2000b) is an algorithm and an interactive tool for the merging two ontologies. The central element of PROMPT is the algorithm which defines a number of steps for the interactive merging process:

1. Identify merge candidates based on class-name similarities. The result is presented to the user as a list of potential merge operations.
2. The user chooses one of the suggested operations from the list or specifies a merge operation directly.
3. The system performs the requested action and automatically executes additional changes derived from the action.
4. The system creates a new list of suggested actions for the user based on the new structure of the ontology, determines conflicts introduced by the last action, finds possible solutions to these conflicts, and displays these to the user.

PROMPT identifies a number of ontology merging operations (merge classes, merge slots, merge bindings between a slot and a class, etc.) and a number of possible conflicts introduced by the application of these operations (name conflicts, dangling references, redundancy in the class hierarchy, and slot-value restrictions that violate class inheritance).

OntoMerge (Dou *et al.*, 2002) is an on-line approach in which source ontologies are maintained after the merge operation, whereas in

PROMPT the merged ontology replaces the source ontologies. The output of the merge operation in OntoMerge is not a complete merged ontology, as in PROMPT, but a bridge ontology which imports the source ontologies and which has a number of Bridging Axioms (see Figure 6.2(b)), which are translation rules used to connect the overlapping part of the source ontologies. The two source ontologies, together with the bridging axioms, are then treated as a single theory by a theorem prover optimized for three main operations:

1. Dataset translation (cf. instance transformation in de Bruijn and Polleres (2004)): dataset translation is the problem of translating a set of data (instances) from one representation to the other.
2. Ontology extension generation: the problem of ontology extension generation is the problem of generating an extension (instance data) O2s, given two related ontologies O1 and O2, and an extension O1s of ontology O1. The example given by the authors is to generate a WSDL extension based on an OWL-S description of the corresponding Web Service.
3. Querying different ontologies: query rewriting is a technique for solving the problem of querying different ontologies, whereas the authors of Dou *et al.* (2002) merely stipulate the problem.

6.3. MAPPING AND QUERYING DISPARATE KNOWLEDGE BASES

In the previous section we have seen an overview of a number of representative approaches for different aspects of ontology mediation in the areas of ontology mapping, alignment, and merging. In this section we focus on an approach for ontology mapping and ontology alignment to query disparate knowledge bases in a knowledge management scenario. However, the techniques are largely applicable to any ontology mapping or alignment scenario.

In the area of knowledge management we assume there are two main tasks to be performed with ontology mappings: (a) transforming data between different representations, when transferring data from one knowledge base to another; and (b) querying of several heterogeneous knowledge bases, which have different ontologies. The ontologies in the area of knowledge management are large, but lightweight, that is, there is a concept hierarchy with many concepts, but there are relatively few relations and axioms in the ontology. From this follows that the mappings between the ontologies will be large as well, and they will generally be lightweight; the mapping will consist mostly of simple correspondence between concepts. The mappings between ontologies are not required to be completely accurate, because of the nature of the application of knowledge management: if a search result is inaccurate it is simply discarded by the user.

In order to achieve ontology mapping, one needs to specify the relationship between the ontologies using some language. A natural candidate to express these relationships would seem to be the ontology language which is used for the ontologies themselves. We see a number of disadvantages to this approach:

- *Ontology language*: there exist several different ontology languages for different purposes (e.g., RDFS (Brickley and Guha, 2004), OWL (Dean and Schreiber, 2004), WSML (de Bruijn *et al.*, 2005)), and it is not immediately clear how to map between ontologies which are specified using different languages.
- *Independence of mapping*: using an existing ontology language would typically require to import one ontology into the other, and specify the relationships between the concepts and relations in the resulting ontology; this is actually a form of ontology merging. The general disadvantage of this approach is that the mapping is tightly coupled with the ontologies; one can essentially not separate the mapping from the ontologies.
- *Epistemological adequacy*: The constructs in an ontology language have not been defined for the purpose of specifying mappings between ontologies. For example, in order to specify the correspondence between two concepts `Human` and `Person` in two ontologies, one could use some equivalence or subclass construct in the ontology language, even though the intension of the concepts in both ontologies is different.

In Section 6.3.1 we describe a mapping language which is independent from the specific ontology language but which can be *grounded* in an ontology language for some specific tasks. The mapping language itself is based on a set of elementary *mapping patterns* which represent the elementary kinds of correspondences one can specify between two ontologies.

As we have seen in Section 6.2.3, there exist many different alignment algorithms for the discovery of correspondences between ontologies. In Section 6.3.2 we present an interactive process for ontology alignment which allows to plug in any existing alignment algorithm. The input of this process consists of the ontologies which are to be mapped and the output is an ontology mapping.

Writing mapping statements directly in the mapping language is a tedious and error-prone process. The mapping tool OntoMap is a graphical tool for creating ontology mappings. This tool described in Section 6.3.3 can be used to create a mapping between two ontologies from scratch or it can be used for the refinement of automatically discovered mappings.

6.3.1. Mapping Language

An important requirement for the mapping language which is presented in this section is the epistemological adequacy of the constructs in the language. In other words, the constructs in the language should correspond to the actual correspondences one needs to express in a natural way. More information about the mapping language can be found in Scharffe and de Bruijn (2005) and on the web site of the mapping language.[1]

Now, what do we mean with 'natural way?' There are different patterns which one can follow when mapping ontologies. One can map a concept to a concept, a concept with a particular attribute value to another concept, a relation to a relation, etc. We have identified a number of such elementary mapping patterns which we have used as a basis for the mapping language.

Example. As a simple example of possible mapping which can be expressed between ontologies, assume we have two ontologies O1 and O2 which both describe humans and their gender. Ontology O1 has a concept `Human` with an attribute `hasGender`; O2 has two concepts `Woman` and `Man`. O1 and O2 use different ways to distinguish the gender of the human; O1 uses an attribute with two possible values '`male`' and '`female`,' whereas O2 has two concepts `Woman` and `Man` to distinguish the gender. Notice that these ontologies have a mismatch in the style of modeling (see Section 6.2.1). If we want to map these ontologies, we need to create two mapping rules: (1) 'all humans with the gender "female" are women' and (2) 'all humans with the gender "male" are men.'

The example illustrates one elementary kind of mapping, namely a mapping between two classes, with a condition on the value of an attribute. The elementary kinds of mappings can be captured in *mapping patterns*. Table 6.1 describes the mapping pattern used in the example.

Table 6.1 Class by attribute mapping pattern.

Name: Class by Attribute Mapping
Problem: The extension of a class in one ontology corresponds to the extension of a class in another ontology, provided that all individuals in the extension have a particular attribute value.

Solution:
Solution description: a mapping is established between a class/attribute/attribute value combination in one ontology and a class in another ontology.
Mapping syntax:

mapping :: = `classMapping(direction A B attributeValueCondition(`Po`))`
Example:
`classMapping(Human Female attributeValueCondition(hasGender`
`'female'))`

---

[1]http://www.omwg.org/TR/d7/d7.2/

The pattern is described in terms of its *name*, the *problem* addressed, the *solution* of the problem, both in natural-language description and in terms of the actual mapping language, and an *example* of the application of the pattern to ontology mapping, in this case a mapping between the class Human in ontology O1 and the class Woman in ontology O2, but only for all humans which have the gender 'female.'

The language contains basic constructs to express mappings between the different entities of two ontologies: from classes to classes, attributes to attributes, instances to instances, but also between any combination of entities like classes to instances, etc. The example in Table 6.1 illustrates the basic construct for mapping classes to classes, classMapping.

Mappings can be refined using a number of operators and mapping conditions. The operators in the language can be used to map between combinations of entities, such as the intersection or union (conjunction, disjunction, respectively) of classes or relations. or example, the mapping between Human and the union of Man and Woman can be expressed in the following way:

```
classMapping(Human or(Man Woman))
```

The example in Table 6.1 illustrates a mapping condition, namely the attribute value condition. Other mapping conditions include attribute type and attribute occurrence.

The mapping language itself is not bound to any particular ontology language. However, there needs to be a way for reasoners to actually use the mapping language for certain tasks, such as querying disparate knowledge bases and data transformation. For this, the mapping language can be grounded in a formal language. There exists, for example, a grounding of the mapping language to OWL DL and to WSML-Flight.

In a sense, the *grounding* of the mapping language to a particular language transforms the mapping language to a language which is specific for mapping ontologies in a specific language. All resulting mapping languages still have the same basic vocabulary for expressing ontology mappings, but have a different vocabulary for the more expressive expressions in the language. Unfortunately, it is not always the case that all constructs in the mapping language can be grounded to the logical language. For example, WSML-Flight does not allow disjunction or negation in the target of a mapping rule and OWL DL does not allow mapping between classes and instances. In order to allow the use of the full expressive power offered by the formal language to which the mapping language is grounded, there is an extension mechanism which allows to insert arbitrary logical expressions inside each mapping rule.

The language presented in this section is suitable for the specification and exchange of ontology mappings. In the next section we present a semi-automatic approach to the specification of ontology mappings.

6.3.2. A (Semi-)Automatic Process for Ontology Alignment

Creating mappings between ontologies is a tedious process, especially if the ontologies are very large. We introduce a semi-automatic alignment process implemented in the Framework for Ontology Alignment and Mapping (FOAM)-tool,[2] which relieves the user of some of the burdens in creating mappings. It subsumes all the alignment approaches we are aware of (e.g., PROMPT (Noy and Musen, 2003), GLUE (Doan *et al.*, 2003), QOM (Ehrig and Staab 2004; Ehrig and Sure 2004)). The input of the process consists of two ontologies which are to be aligned; the output is a set of correspondences between entities in the ontologies. Figure 6.3 illustrates its six main steps.

1. *Feature engineering*: it selects only parts of an ontology definition in order to describe a specific entity. For instance, alignment of entities may be based only on a subset of all RDFS primitives in the ontology. A feature may be as simple as the label of an entity, or it may include intentional structural descriptions such as super- or sub-concepts for concepts (a sports car being a subconcept of car), or domain and range for relations. Instance features may be instantiated attributes. Further, we use extensional descriptions. In an example we have fragments of two different ontologies, one describing the instance Daimler and one describing Mercedes. Both o1:Daimler and o2:Mercedes have a generic ontology feature called type. The values of this feature are automobile and luxury, and automobile, respectively.

2. *Selection of next search steps*: next, the derivation of ontology alignments takes place in a search space of candidate pairs. This step may choose to compute the similarity of a restricted subset of candidate concepts pairs of the two ontologies and to ignore others. For the running example we simply select every possible entity pair as an alignment candidate. In our example this means we will continue the comparison of o1:Daimler and o2:Mercedes. The QOM approach of Section 6.2.3 carries out a more efficient selection.

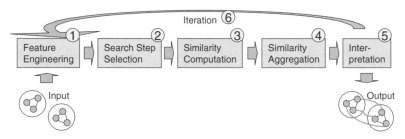

Figure 6.3 Alignment process.

[2]http://www.aifb.uni-karlsruhe.de/WBS/meh/foam

Table 6.2 Feature/similarity assessment.

Comparing	No.	Feature Q_F	Similarity Q_S
Entities	FS1	(label, X_1)	string similarity (X_1, X_2)
Instances	FS2	(parent, X_1)	set equality (X_1, X_2)

3. *Similarity assessment*: it determines similarity values of candidate pairs. We need heuristic ways for comparing objects, that is similarity functions such as on strings, object sets, checks for inclusion, or inequality, rather than exact logical identity. In our example we use a similarity function based on the instantiated results, that is we check whether the two concept sets, parent concepts of o1:Daimler (automobile and luxury), and parent concepts of o2:Mercedes (only automobile) are the same. In the given case this is true to a certain degree, effectively returning a similarity value of 0.5. The corresponding feature/similarity assessment (FS2) is represented in Table 6.2 together with a second feature/similarity assessment (FS1) based on the similarity of labels.

4. *Similarity aggregation*: in general, there may be several similarity values for a candidate pair of entities from two ontologies, for example one for the similarity of their labels and one for the similarity of their relationship to other terms. These different similarity values for one candidate pair must be aggregated into a single aggregated similarity value. This may be achieved through a simple averaging step, but also through complex aggregation functions using weighting schemes. For example, we only have to result of the parent concept comparison which leads to: simil(o1:Daimler,o2:Mercedes) = 0.5.

5. *Interpretation*: it uses the aggregated similarity values to align entities. Some mechanisms here are, for example to use thresholds for similarity (Noy and Musen, 2003), to perform relaxation labeling (Doan *et al.*, 2003), or to combine structural and similarity criteria. simil(o1:Daimler,o2:Mercedes) = 0.5 ≥ 0.5 leads to align(o1:Daimler) = o2:Mercedes. This step is often also referred to as *matcher*. Semi-automatic approaches may present the entities and the alignment confidence to the user and let the user decide.

6. *Iteration*: several algorithms perform an iteration (see also similarity flooding (Melnik *et al.*, 2002)) over the whole process in order to bootstrap the amount of structural knowledge. Iteration may stop when no new alignments are proposed, or if a predefined number of iterations has been reached. Note that in a subsequent iteration one or several of steps 1 through 5 may be skipped, because all features might already be available in the appropriate format or because some similarity computation might only be required in the first round. We use the intermediate results of step 5 and feed them again into the process and stop after a predefined number of iterations.

The output of the alignment process is a mapping between the two input ontologies. We cannot in general assume that all mappings between the ontologies are discovered, especially in the case of more complex mappings. Therefore, the mapping which is a result of the alignment procedure can be seen as the input of a manual refinement process. In the next section we describe a graphical tool which can be used for manual editing of ontology mappings.

6.3.3. OntoMap: an Ontology Mapping Tool

OntoMap® (Schnurr and Angele, 2005; see Figure 6.4) is a plugin for the ontology-management platform OntoStudio® that supports the creation and management of ontology mappings. Mappings can be specified based on graphical representation, using a schema-view of the respective ontologies. OntoMap encapsulates the formal statements for the declaration of mappings, users only need to understand the semantics of the graphical representation (e.g., an arrow connecting two concepts). Users of OntoMap are supported by drag-and-drop functionality and simple consistency checks on property mappings (automatic suggestion of necessary class mappings). For concept to concept mappings constraints can be specified on the available attributes.

OntoMap supports a number of most elementary mapping patterns: concept to concept mappings, attribute to attribute mappings, relation to relation mappings, and attribute to concept mappings.

Additionally, OntoMap allows to specify additional conditions on concept to concept mappings using a form for mapping properties. A concept 'lorry' for example might map onto a concept 'Car or truck' only if the weight of the latter exceeds a certain limit (e.g., according to the legal definition within some countries).

Attribute to concept mappings enable users to specify one or more 'identifiers' for instances of a concept—similar to the primary keys in relational databases. This way the properties of different source concepts can be 'unified' in one target concept, for example in order to join the information of different database entries within a single instance on the ontology level. Different source instances having the same 'identifier values' are then joined within a single target instance.

The focus of OntoMap is on the intuitive creation and management of mappings. If complex mappings are needed, which are not within the scope of a graphical tool (possibly using complex logical expressions or built-ins), they have to be encoded manually. OntoStudio has its own grounding of mappings, based on F-Logic rules (Kifer and Lausen, 1997). In addition to the internal storage format OntoMap supports the import and export of mappings in the mapping language which we described in this section. An extension of OntoMap based on a library for ontology alignment which was described in Section 6.3.2 provides the functionality for the semi-automatic creation of mappings.

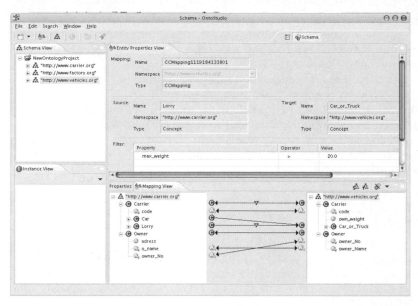

Figure 6.4 Screenshot of OntoStudio® with the OntoMap® plugin.

Some additional features of the OntoStudio environment support users in typical mediation tasks. Those include the import of schemas for relational databases (syntactic integration) and the possibility to create and execute queries instantly. The latter gives users the possibility to test the consequence of mappings they have created (or imported).

6.4. SUMMARY

Overlap and mismatches between ontologies are likely to occur when the vision of a Semantic Web with a multitude of ontologies becomes a reality.

There exist different approaches to ontology mediation. These approaches can be broadly categorized as: (a) *Ontology Mapping* (Maedche *et al.*, 2002; Scharffe and de Bruijn, 2005), (b) *Ontology Alignment* (Rahm and Bernstein, 2001; Doan *et al.*, 2004; Ehrig and Staab, 2004; Ehrig and Sure, 2004), and (c) *Ontology Merging* ((Noy and Musen, 2000b; Dou *et al.* 2002)). We have presented a survey of the most prominent approaches in these areas.

Additionally, we described a practical approach to representing mappings using a mapping language, discovering mappings using an alignment process which can be used in combination with any ontology alignment algorithm, and a graphical tool to edit such ontology mappings. These ontology mappings can be used, for example, for transforming data between different representation, as well as querying different heterogeneous knowledge bases.

Although there is some experience with ontology mediation and most approaches to ontology mediation, especially in the field of ontology alignment, have been validated using some small test set of ontologies, the overall problem which the area of ontology mediation faces is that the number of ontologies which is available on the Semantic Web is currently very limited, and it is hard to validate the approaches using real ontologies.

REFERENCES

Bouquet P, Giunchiglia F, van Harmelen F, Serafini L, Stuckenschmidt H. 2004. Contextualizing ontologies. *Journal of Web Semantics* 1(4):325.

Brickley D, Guha RV. 2004. RDF vocabulary description language 1.0: RDF schema, W3c recommendation 10 February 2004, W3C. URL: http://www.w3.org/TR/rdf-schema/

de Bruijn J, Fensel D, Keller U, Kifer M, Lausen H, Krummenacher R, Polleres A, Predoiu L. 2005. The web service modeling language WSML, W3C member submission 3 June 2005. URL: http://www.w3.org/Submission/WSML/

de Bruijn J, Polleres A. 2004. Towards and ontology mapping specification language for the semantic web, Technical Report DERI-2004-06-30, DERI. URL: http://www.deri.at/publications/techpapers/documents/DERI-TR-2004-06-30.pdf

Dean M, Schreiber G. 2004. OWL web ontology language reference, W3C recommendation 10 February 2004. URL: http://www.w3.org/TR/owl-ref/

Doan A, Domingos P, Halevy A. 2003. Learning to match the schemas of data sources: A multistrategy approach. *VLDB Journal* 50:279–301.

Doan A, Halevy A. 2005. Semantic integration research in the database community: A brief survey, AI Magazine, Special Issue on Semantic Integration.

Doan A, Madhaven J, Domingos P, Halevy A. 2004. Ontology matching: A machine learning approach. *Handbook on Ontologies in Information Systems*, In Staab S, Studer R (eds). Springer-Verlag, pp 397–416.

Dou D, McDermott D, Qi P. 2002. Ontology translation by ontology merging and automated reasoning, In *'Proceedings EKAW2002 Workshop on Ontologies for Multi-Agent Systems'*, pp 3–18.

Ehrig M, Staab S. 2004. QOM—quick ontology mapping. In *Proceedings of the Third International SemanticWeb Conference (ISWC2004)*, van Harmelen F, McIlraith S, Plexousakis D (eds). LNCS, Springer: Hiroshima, Japan, pp 683–696.

Ehrig M, Sure Y. 2004. Ontology mapping—an integrated approach. In *Proceedings of the First European Semantic Web Symposium, ESWS* 2004, Vol. 3053 of Lecture Notes in Computer Science, Springer-Verlag, Heraklion, Greece, pp 76–91.

Giunchiglia F, Shvaiko P. 2004. Semantic matching. *The Knowledge Engineering Review* 18(3):265–280.

Giunchiglia F, Shvaiko P, Yatskevich M. 2004. S-match: An algorithm and an implementation of semantic matching. In *Proceedings of ESWS'04*, number 3053 in 'LNCS', Springer-Verlag, Heraklion, Greece, pp 61–75.

Kalfoglou Y, Schorlemmer M. 2003. Ontology mapping: The state of the art. *The Knowledge Engineering Review* 18(1):1–31.

Kifer M, Lausen G. 1997. F-logic: A higher-order language for reasoning about objects. *SIGMOD Record* 18(6):134–146.

Klein M. 2001. Combining and relating ontologies: An analysis of problems and solutions. In *'Workshop on Ontologies and Information Sharing*, Gomez-Perez A, Gruninger M, Stuckenschmidt H, Uschold M (eds). IJCAI'01' Seattle, USA.

Maedche A, Motik B, Silva N, Volz R. 2002. Mafra—a mapping framework for distributed ontologies. In *Proceedings of the 13th European Conference on Knowledge Engineering and Knowledge Management* EKAW-2002, Madrid, Spain.

Melnik S, Garcia-Molina H, Rahm E. 2002. Similarity flooding: A versatile graph matching algorithm and its application to schema matching. In *Proceedings of the 18th International Conference on Data Engineering (ICDE'02)*, IEEE Computer Society, 117 p.

Noy NF. 2004. Semantic integration: A survey of ontology-based approaches. Sigmod Record, Special Issue on Semantic Integration 33(4):65–70.

Noy NF, Musen MA. 2000a. Anchor-PROMPT: Using non-local context for semantic matching. In *Proceedings of the Workshop on Ontologies and Information Sharing at the Seventeenth International Joint Conference on Artificial Intelligence* (IJCAI-2001), Seattle, WA, USA.

Noy NF, Musen MA. 2000b. PROMPT: Algorithm and tool for automated ontology merging and alignment. In *'Proceedings 17th National Conference On Artificial Intelligence (AAAI2000)'*, Austin, Texas, USA.

Noy NF, Musen MA. 2003. The PROMPT suite: Interactive tools for ontology merging and mapping. *International Journal of Human-Computer Studies* 59(6): 983–1024.

Omelayenko B. 2002. RDFT: A mapping meta-ontology for business integration. In *Proceedings of the Workshop on Knowledge Transformation for the Semantic Web (KTSW 2002) at the 15th European Conference on Artificial Intelligence*, Lyon, France, pp 76–83.

Rahm E, Bernstein PA. 2001. A survey of approaches to automatic schema matching. *VLDB Journal: Very Large Data Bases* 10(4):334–350.

Scharffe F, de Bruijn J. 2005. A language to specify mappings between ontologies. In *Proceedings of the Internet Based Systems IEEE Conference (SITIS05)*, Yandoue, Cameroon.

Schnurr HP, Angele J. 2005. Do not use this gear with a switching lever! automotive industry experience with semantic guides. In 4th International Semantic Web Conference (ISWC2005), pp 1029–1040.

Shvaiko P, Euzenat J. 2005. A survey of schema-based matching approaches. *Journal on Data Semantics* 4:146–171.

Uschold M. 2000. Creating, integrating, and maintaining local and global ontologies. In *Proceedings of the First Workshop on Ontology Learning (OL-2000) in conjunction with the 14th European Conference on Artificial Intelligence (ECAI-2000)*, Berlin, Germany.

Visser PRS, Cui Z. 1998. On accepting heterogeneous ontologies in distributed architectures. In *Proceedings of the ECAI98 workshop on applications of ontologies and problem-solving methods*, Brighton, UK.

7

Ontologies for Knowledge Management

Atanas Kiryakov

7.1. INTRODUCTION

This chapter discusses a number of aspects of the usage of ontologies in Knowledge Management (KM) context, as well as in some specific Semantic Web applications. The semantic annotation of unstructured content, that is linking it to ontologies, is relevant to the subject, but not addressed in detail here—annotation and its usage for indexing, hyper-linking, visualization, and navigation is discussed in Chapter 3.

We start with a simple motivating scenario, showing some benefits of using ontologies in database-like setting. Next, we clarify the meaning and typical usage of some terms related to ontologies (e.g., taxonomy, knowledge base, metadata) and discuss different types of ontology: upper-level versus domain-specific; lightweight versus heavyweight, and so on.

Two of the possible roles of ontologies are discussed: as a database schema and as a topic hierarchy. The different requirements and restrictions specific to the ontologies used in these roles are commented.

The remainder of the chapter presents the PROTON ontology, as an example for an ontology designed to support a number of KM and Semantic Web applications. Some concrete ontology design examples and recommendations are given in this context.

Semantic Web Technologies: Trends and Research in Ontology-based Systems
John Davies, Rudi Studer, Paul Warren   © 2006 John Wiley & Sons, Ltd

7.2. ONTOLOGY USAGE SCENARIO

Formal knowledge representation (KR) is about building models[1] of the world, of a particular domain or a problem, which allow for automatic reasoning and interpretation. Such formal models are called ontologies and can be used to provide formal semantics (i.e., machine-interpretable meaning) to any sort of information: databases, catalogs, documents, web pages, etc. The association of information with such formal models makes the information much amenable to machine processing and interpretation.

Semantic repositories[2] allow for storage, querying, and management of structured data. They can serve as a replacement for database management systems (DBMS), offering easier integration of diverse data and more analytical power. In a nutshell, a semantic repository can dynamically interpret metadata schemata and ontologies, which determine the structure and the semantics of data and of queries against that data.

Compared to the approach taken in relational DBMSs, this allows for (i) easier changes to and combinations of data schemata and (ii) automated interpretation of the data.

As an example, let us imagine a typical database populated with the information that John is a son of Mary. It will be able to 'answer' just a couple of questions: *Which are the son(s) of Mary?* and *Of whom is John the son?* Given simple family relationships ontology (as the one presented in Figure 7.1), a semantic repository could handle much bigger set of

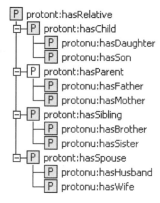

Figure 7.1 Hierarchy of family relations.

---

[1]The typical modeling paradigm is mathematical logic, but there are also other approaches, rooted in information and library science. KR is a very broad term; here we only refer to one of its main streams.

[2]'Semantic repository' is not a well-established term. See http://www.ontotext.com/inference/semantic_repository.html for a more elaborate introduction.

questions. It will be able infer the more general fact that John is a child of Mary and, even more generally, that Mary and John are relatives (which is true in both directions because `hasRelative` is defined to be symmetric in the ontology). Further, if it is known that Mary is a woman, a semantic repository will infer that Mary is the mother of John, which is a more specific inverse relation. Although simple for a human to infer, the above facts would remain unknown to a typical DBMS and indeed to any other information system, for which the model of the world is limited to datastructures of strings and numbers with no automatically interpretable semantics.

7.3. TERMINOLOGY

We provide here definitions of terms related to ontologies and their usage in the KM and Semantic Web context.

Dublin Core Metadata Initiative (*DCMI* or *DC*) is an 'open forum engaged in the development of interoperable online metadata standards that support a broad range of purposes and business models,' (DCMI, 2005). DC is used here mostly for the sake of reference to a widely shared terminology.

Dataset is any set of structured data, as defined in DC, (DCMI, 2003b): 'A dataset is information encoded in a defined structure (for example, lists, tables, and databases), intended to be useful for direct machine processing.'

Ontology is a term having different meaning in the disciplines of Philosophy and Computer Science (CS). It was originally defined by philosophers as a discipline concerned with the nature of being and existence.

In the field of CS (and IT in general) it has become popular as a paradigm for knowledge representation in Artificial Intelligence (AI), by providing a methodology for easier development of interoperable and reusable knowledge bases. The most popular definition, from an AI perspective, is given in Gruber (1992) as follows: 'An ontology is an explicit specification of a conceptualization,' where 'a conceptualization is an abstract, simplified view of the world that we wish to represent for some purpose.' Another widely used extended definition is provided in Borst (1997): 'An ontology is a formal, explicit specification of a shared conceptualization.' An extended discussion on the terminology is provided in Gruber (1992).

Here we would like to add that ontologies can be considered as conceptual schemata, intended to represent knowledge in the most formal and reusable way possible. Formal ontologies are represented in logical formalisms, such as OWL (Dean *et al.*, 2004), which allow automatic inferencing over them and over datasets aligned to them. An important role of ontologies is to serve as schemata or 'intelligent' views

over information resources.[3] Thus they can be used for indexing, querying, and reference purposes over nonontological datasets and systems,
such as databases, document and catalog management systems. Because
ontological languages have a formal semantics, ontologies allow a wider
interpretation of data, that is inference of facts which are not explicitly
stated. In this way, they can improve the interoperability and the
efficiency of the usage of arbitrary datasets.

Ontologies are typically classified depending on the generality of the
conceptualization behind them, their coverage, and intended purpose:

- *Upper-level ontologies* represent a general model of the world, suitable
  for large variety of tasks, domains, and application areas.
- *Domain ontologies* represent a conceptualization of a specific domain,
  for example road-construction or medicine.
- *Application and task ontologies* are such suitable for specific ranges of
  applications and tasks. An example of such is the PROTON KM
  module (see Subsection 7.6.4).

A more extensive overview of the different sorts of ontologies and their
usage can be found in Guarino (1998b), which also provides discussion
on the different ways in which 'ontology' is used as a term and its
relation to knowledge bases.

Knowledge Base (KB) is a term with a wide usage and multiple meanings. Here we consider a KB as a dataset with some formal semantics. A
KB, similarly to an ontology, is represented with respect to a knowledge
representation formalism, which allows automatic inference. It could
include multiple axioms, definitions, rules, facts, statements, and any
other primitives. In contrast to ontologies, KBs are not intended to
represent a (shared/consensual) schema, a basic theory, or a conceptualization of a domain. Thus, ontologies are a specific sort of knowledge
base. An ontology can be characterized as comprising a 4-tuple:[4]

$$O = \langle C, R, I, A \rangle$$

Where C is a set of classes representing *concepts* we wish to reason about
in the given domain (invoices, payments, products, prices,...); R is a set
of *relations* holding between those classes (`Product hasPrice Price`); I
is a set of *instances*, where each instance can be an instance of one or more
classes and can be linked to other instances by relations (`product17
isA Product; product23 hasPrice €170`); A is a set of *axioms* (if a
product has a price greater than €200, shipping is free).

---

[3]Comments in the same spirit are provided in Gruber (1992) also. This is also the role of
ontologies in the Semantic Web.
[4]Note that a more formal and extensive mathematical definition of an ontology is given in
Chapter 2. The characterization offered here is suitable for the purposes of our discussion,
however.

It is widely recommended that knowledge bases, containing concrete data[5] are always encoded with respect to ontologies, which encapsulate a general conceptual model of some domain knowledge, thus allowing easier sharing and reuse of KBs.

Typically, ontologies designed to serve as schema[6] for KBs do not contain instance definitions, but there is no formal restriction in this direction. Drawing the borderline between the ontology (i.e., the conceptual and consensual part of the knowledge) and the rest of the data, represented in the same formal language, is not always a trivial task. For instance, there could be an ontology about tourism, which defines the classes `Location` and `Hotel`, as well as the `locatedIn` relation between them and the hotel attribute `category`. The definitions of the classes, relations, and attributes should clearly be a part of the ontology. The information about a particular hotel is probably not a part of the ontology, as far as it is not a matter of conceptualization and consensus, but is just a description, crafted for some (potentially specific) purpose. Then, suppose that there is a definition of `New York` as an instance of the class `City`—it can be argued that it is either a part of the ontology or just a description of a city. The fact that it is an instance does not necessarily determine that it is not part of the conceptualization.

Let us assume that a knowledge engineer is guided by the principle 'no instances in ontologies.' Even in this case there are many examples when one and the same concept can be represented as both class and instance, so, this design principle does not help us always to determine what should be part of a schema-ontology, and what not. As an example, 'VW Golf' (as a model) can be an instance of 'VW Car.' However, it also make sense to define a specific vehicle (e.g., golf-12643789) of this model as an instance of 'VW Golf' (taken as a class). There is no simple way to determine whether 'VW Golf' should be defined as class or instance in this case—such modeling decisions are to some extent a function of the intended use of the ontology.

7.3.1. Data Qualia

Below we present a few boolean qualia[7] of the data relevant to the ontology representation and data integration problems:[8]

---

[5]Often referred as instance data, instance knowledge, A-Box, etc.

[6]Notice that the term ontology has become somewhat overloaded and ambiguous in recent years in the Computer Science community. There are many authors which use ontology as a place holder for any sort of KB and even any sort of conceptual model, including such without formal semantic. We find such interpretations ambiguous and confusing and stick to the 'classical' definition here.

[7]Quale (pl. Qualia), is here used as a primary intrinsic quality, an independent (orthogonal to others) dimension of classification. According to the Merriam-Webster online dictionary (1) a property (as redness) considered apart from things having the property, UNIVERSAL; (2) a property as it is experienced as distinct from any source it might have in a physical object.

[8]This analysis of the different sorts of data was first published in Kiryakov (2004b).

- *Semantics*: whether the semantics (the meaning) of the data is formally represented, so that a machine can formally interpret it, reason and derive new data?[9] This quale is directly relevant to reasoning and ontology management—reasoning can only be performed on top of 'semantic' data. Nonsemantic data could be adapted for reasoning by means of mapping it to an ontology, that is a semantic schema which defines the meaning of the data externally. There are marginal cases where the specification of a structure bears elements of semantics, for example the case of XML schemata. We stay with a relatively narrow definition of what semantics are and consider semantic data only when there is some logical theory defining meaning associated with the representation language used to represent or interpret the data.
- *Structure*: whether the data is formally structured, so that a machine can formally interpret and manage its structure? This distinction is important because the approaches for automated access and management (and their typical performance) differ considerably between structured and unstructured data.
- *Schema*: here we consider schematic data, which determines the structure and/or the semantics of other data. Obviously, there are schematic and nonschematic data. The schema quale is determined by the (intended) role of the data with respect to other data. This distinction is relevant within the ontology management context for the following reasons:

  - Schemata are important for mediation and evolution because these determine the consistency and the interpretation of other data. For instance, a change in an ontology can render a dataset previously compliant with the old version, incompliant with the new one (or vice versa).
  - In many cases, the problem of data integration can be solved at the level of schema integration.

7.3.2. Sorts of Data

We introduce a short analysis of the different sorts of data available, distinguished with respect to the qualia presented in the previous section (semantics, structure, and schema). The analysis facilitates the further discussion of different sorts of ontologies and their roles. The analysis follows (the values for the three qualia are given in brackets, where _ stands for 'any value'):

- *Data*, (_,_,_). Any sort of data.
  - Datasets, (_,structured,_). See the definition above, referring to Dublin Core.

---

[9]The newly inferred data are expected to be correct, indisputable from the human perspective, and a consequence of the explicit data.

- Knowledge Bases, (`semantic,structured,_`). Any sort of a dataset with a well-defined formal semantics. Those are often referred to as instance datasets or instance knowledge. See the definition in the previous subsection.
  - Ontologies, (`semantic,structured,schema`). See the definition in the previous subsection. Ontologies are used to prescribe both structure and semantics. For instance, an ontology can define the valid attributes for a specific class (like a database schema can do, too) and, in addition, it can specify the semantics of the attributes.
- Nonsemantic schemata, (`nonsemantic,structured,sch-ema`). Such examples are database and XML schemata.
- Databases, (`nonsemantic, structured, nonschema`). Here databases are used as a generic term for relational databases, XML-encoded data, comma-separated files, and any other structured, nonsemantic data that is not intended to serve as a schema, but rather to represent or communicate particular information. Although this is a slightly misleading name, it reflects the fact that relational databases are the most important sort of nonsemantic, nonschema data.
- Mixed datasets, (`_,structured,schema&non-schema`). Many catalogs and taxonomies can serve as examples. In such datasets one can often find subsumption chains of the sort Location-City-New York, with no formal indication that the first two are classses (schema) and the third is an instnance (nonschema).

- Content, (`_,non-structured,_`). Any data without a substantial machine-understandable structure. Such examples are free-text documents, pictures, voice or video recordings, etc. In most of these cases, the non-structured data neither bears machine-interpretable semantics nor plays the role of a formal schema.

Metadata is a term of a wide and often controversial or misleading usage. From its etymology, metadata is 'data about data.' Thus, metadata is a role that certain data could play with respect to other data. Such an example could be a particular (structurally) formal specification of the author of a document, provided independently from the content of the document, say, according to a standard like DC. RDF(S), (Klyne and 2004; Caroll Brickley and Guha, 2000), has been introduced as a simple KR language that is to be used for the assignment of semantic descriptions to information resources on the web. Therefore an RDF description of a web page represents metadata. However, an RDF description of a person, independent from any particular documents (e.g., as a part of an RDF(S)-encoded dataset), is not metadata—this is data about a person, not about other data. In the latter case, RDF(S) is used as a KR language. Finally, the RDF(S) definition of the class Person, should typically be part of an ontology, which can be used to structure datasets and metadata, but which is again not a piece of metadata itself. A term, which is often used as a synonym for metadata, is *annotation*. However, it also has a special meaning in the natural language processing (NLP) community. Please refer to Chapter 3 for a discussion on 'semantic annotation.'

Metadata is another candidate for an information quale (in addition to the three presented above). However, it is not presented this way because we regard the term as more representing a role for the data rather than a quality.[10]

Semi-structured data is a term used to refer to two different notions. First, in the KM and NLP communities, semi-structured data are usually considered documents that contain free-text fragments, structured in accordance with some schema. Typical sorts of semi-structured documents are forms and tables, which have some strict structure (fields, parts, etc.), whilst the content of the specific parts of the document is a free text. Examples are many administrative, insurance, customs, and medical forms. The second usage of the term 'semi-structured' is rather different, denoting nonrelational data models (Figure 7.2). The intuition is that, whilst with databases there is a predefined, strict structure of specific tables, fields, and views, there are other, 'semi-structured,' representations with less strict structuring, which are still not unstructured.[11] A number of, more or less, graph-based data-models, like RDF

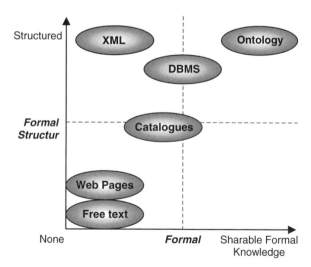

Figure 7.2 Structured versus semantic positioning of different sorts of data.

---

[10]This is also the case with the Schema quale, but to a smaller degree, in our opinion.

[11]See Subsection 3.1.2 of Martin-Recuerda *et al.* (2004) for extended discussion on semi-structured data and its relation to Object Exchange Model (OEM).

and the Associative Data Model, described in Williams (2002), match this understanding of semi-structured data. In both cases, there are two levels of structuring. At the logical[12] level, there is a very simple model, which can be used as a general carrier or canvas for the representation of the data. On top of it, there could be a 'softer' and much more dynamic schema, which supports the interpretation of the data stored in the basic model. If we take the latter meaning of 'semi-structured,' RDFS and OWL are semi-structured representations. However, we strongly disagree with the philosophy behind this usage of semi-structured. Languages and models like RDF(S) allow dynamic and flexible structuring, which, in our view, is a higher degree of structuring, instead of a 'semi'-one. Thus, further in this chapter, we will only use semi-structured as a term for (text) documents with partial structure (i.e., the first meaning).

7.4. ONTOLOGIES AS RDBMS SCHEMA

Here we discuss formal ontologies modeled through KR formalisms based on mathematical logic (ML); there is a note on so-called topic-ontologies in a subsection below. If we compare ontologies with the schemata of the relational DBMS, there is no doubt that the former represent (or allow for representations of) richer models of the world. There is also no doubt that the interpretation of data with respect to the fragments of ML which are typically used in KR is computationally much more expensive as compared to interpretation using a model based on relational algebra. From this perspective, the usage of the lightest possible KR approach, that is the least expressive (but still adequate) logical fragment, is critical for the applicability of ontologies in more of the contexts in which DBMS are used.

In our view, what is very important for the DBMS paradigm is that it allows for management of huge amounts of data in a predictable and verifiable fashion. It is relatively easy to understand a relational database schema: most computer science (CS) graduates would have a good grasp of the concepts involved. We can assume that the efforts for under-standing and management of such a schema grow in an approximately linear way with its size. Again, someone with a general CS background can predict, understand, and verify the results of a query, even on top of datasets with millions or billons of records. This is the level of control and manageability required for systems managing important data in enterprises and public service organizations. And this is the requirement which is not well covered by the heavyweight, fully fledged, logically expressive knowledge engineering approaches. Even taking a trained knowledge engineer and a relatively simple logical fragment (e.g., OWL DL),

---

[12]With regard to the database terminology.

it is significantly more complex for the engineer to maintain and manage the ontology and the data, as the size of the ontology and the scale of the data increase. We leave the above statements without proof, anticipating that most of the readers either share our observations and intuition[13] or are prepared to take them on trust.

Ontologies can be informally divided into *lightweight* and *heavyweight* according to the expressivity of the KR language used for their formalization and the basic modeling and design principles enforced. Heavyweight (also sometimes referred to as fully fledged) ontologies usually provide complete definitions (of classes, properties, etc.), but fail to match the scalability and manageability requirements for the database-schema-replacement scenario. Lightweight ontologies are usually less restrictive. In other words, the conceptualization behind them is a more general one; the definitions are rather partial; the possible interpretations are not constrained to the degree possible for heavyweight ontologies. This limits the 'predictive' (or the restrictive) power of lightweight ontologies. Often upper-level ontologies are lightweight because without domain constraints it proves hard to craft universal and consensual complete definitions.

7.5. TOPIC-ONTOLOGIES VERSUS SCHEMA-ONTOLOGIES

There is a wide range of applications for which the classification of different things (entities, files, web-pages, etc.) with respect to hierarchies of topics, subjects, categories, or designators has proven to be a good organizational practice, which allows for efficient management, indexing, storage, or retrieval. Probably the most well-known example in this area are library classification systems. Another is given by taxonomies, which are widely used in the KM field. Finally, Yahoo and DMoz[14] are popular and very large scale incarnations of this approach in the context of the World Wide Web. A number of the most popular taxonomies are listed as encoding schemata in Dublin Core, Section 4 in (DCMI, 2005).

Given that the above-mentioned conceptual hierarchies represent a form of shared conceptualization, it is not surprising that they are often considered as ontologies of some kind. It is our view, however, that these ontologies bear a different sort of semantics. The formal framework, which allows for efficient interpretation of DB-schema-like ontologies (such as PROTON, which we discuss in more detail in Section 7.6), is not

---

[13]We are tempted to share a hypothesis regarding the source of the unmanageability of any reasonably complex logical theory. It is our understanding that Mathematical Logic provides a rough approximation for the process of human thinking, but one which renders it hard to follow. Relational algebra is also a rough approximation, but it seems simple enough to be understood by a trained person.

[14]http://www.yahoo.com and http://www.dmoz.org, respectively.

that suitable and compatible with the semantics of topic hierarchies. For the sake of clarity, we introduce the terms 'schema-ontology' and 'topic-ontology.'

To provide a better understanding of the distinctions between topic- and schema-ontologies, we will briefly sketch the formal modeling of the semantics of the latter. Schema-ontologies are typically formalized with respect to so-called *extensional semantics*, which in its simplest form allows for a two-layered set-theoretic model of the meaning of the schema elements. It can be briefly characterized as follows:

- The set of classes and relations on one hand is disjoint from the set of individuals (or instances), on the other. These two sets form the vocabu- laries, respectively, of the TBox and the ABox in description logics.
- The semantics of classes are defined through the sets of their instances. Namely, the interpretation of a class is the set of its instances. The sub- class operation in this case is modeled as set inclusion (as in classical algebraic set theory).
- Relations are defined through the sets of ordered n-tuples (the sequences of parameters or arguments) for which they hold. Sub- relations are again defined through sub-sets. In the case of RDF/OWL properties, which are binary relations, their semantics are defined as sets of ordered pairs of subjects and objects.
- This model can easily be extended to provide a mathematical ground- ing for various logical and KR operators and primitives, such as cardinality constraints.
- Everything which cannot be modeled through set inclusion, member- ship, or cardinality within this model is indistinguishable or 'invisible' for this sort of semantics—it is not part of the way in which the symbols are interpreted.

The computational efficiency of languages with extensional semantics (in terms of induction and deduction algorithms) is well understood. Typi- cal and interesting examples are the family of description logics, and in particular OWL DL and the other OWL species where the trade-off between expressivity and computational tractability have been well explored.[15]

The semantics of topics have a different nature. Topics can hardly be modeled with set-theoretic operations—their semantics have more in common with so-called intensional semantics. In essence, the distinction is that the semantics are not determined by the set of instances (the extension), but rather by the definition itself and more precisely the information content of the definition. Intensional semantics are in a sense closer to the associative thinking of the human being than ML (in its simple incarnations). The criteria for whether a topic is a sub-topic of

---

[15]http://www.w3.org/TR/owl-guide/

another topic do not have much to do with the sets of instances of the respective class (if topics are modelled as classes). To some extent this is because the notion of 'being an instance' is hard to define in this context.

Even disregarding the hypothesis for the different nature of the semantics of the topic- and schema-ontologies, we suggest that these should be kept detached. The hierarchy of classes of the latter should not be mixed up with topic hierarchies because this can easily generate paradoxes and inconsistent ontologies. Imagine, for example, a schema-ontology, where we have definitions for `Africa` and `AfricanLion`[16]— it is likely that `Africa` will be an instance of the `Continent` class and `AfricanLion` will be a sub-class of `Lion`. Imagine also a book classification—in this context `AfricanLionSubject` can be subsumed by `AfricaSubject` (i.e., books about AfricanLions are also about Africa). If we had tried to 'reuse' for classification purposes the definitions of `Africa` and `AfricanLion` from the schema-ontology, this would require that we define `AfricanLion` as a sub-class of `Africa`. The problems are obvious: `Africa` is not a class, and there is no easy way to redefine it so that the schema-ontology extensional sub-classing coincides with the relation required in the topic hierarchy. This example was proposed by the authors, to Natasha Noy for the sake of support of Approach 3 within the ontology modeling study published in Noy (2004). One can find there some further analysis on the computational complexity implications of different approaches to the modeling of topic hierarchies.

7.6. PROTON ONTOLOGY

The PROTON (PROTo ONtology) ontology has been developed in the SEKT project as a lightweight upper-level ontology, serving as the modeling basis for a number of tasks in different domains. To mention just a few applications: PROTON is meant to serve as a seed for ontology generation (new ontologies constructed by extending PROTON); it is further used for automatic entity recognition and more generally Information Extraction (IE) from text, for the sake of semantic annotation (metadata generation).

7.6.1. Design Rationales

PROTON is designed as a lightweight upper-level ontology for usage in Knowledge Management and Semantic Web applications. The above mission statement has two important implications:

---

[16]The example would perhaps have been more intuitive if we had use AfricanTribes instead of AfricanLion, but we prefer to use the same classes and topics as the example given in Noy (2004).

- PROTON is relatively unrestrictive (see the comments on lightweight ontologies above).
- PROTON is naïve in some aspects, for instance regarding the conceptualization of space and time. This is partly because proper models for these aspects would require usage of logical apparatus which is beyond the limits acceptable for many of the tasks to which we wish to apply PROTON (e.g., queries and management of huge datasets/knowledge bases); and partly because it is very hard to craft strict and precise conceptualizations for these concepts which are adequate for a wide range of domains and applications.

Having accepted the above drawbacks, we add two additional requirements to PROTON; namely, to allow for (i) low cost of adoption and maintenance and (ii) scalable reasoning. The goal is to make feasible the usage of ontologies and the related reasoning infrastructure (with all their attendant advantages discussed above) as a replacement for the use of DBMSs.

Being lightweight, PROTON matches the intuition behind the arguments coming from the Information Science community, (Sparck Jones, 2004; Shirky, 2005), that the Semantic Web is more likely to yield solutions to real world information management problems if it is based on partial and relatively simple models of the world, used for semantic tagging.

7.6.2. Basic Structure

The PROTON ontology contains about 300 classes and 100 properties, providing coverage of the general concepts necessary for a wide range of tasks, including semantic annotation, indexing, and retrieval. The design principles can be summarized as follows (i) domain-independence; (ii) lightweight logical definitions; (iii) alignment with popular metadata standards; (iv) good coverage of named entity types and concrete domains (i.e., modeling of concepts such as people, organizations, locations, numbers, dates, addresses, etc.). The ontology is encoded in a fragment of OWL Lite and split into four modules: System, Top, Upper, and Knowledge Management (KM). A snapshot of the PROTON class hierarchy is given in Figure 7.3, showing the Top and the Upper modules.

PROTON is presented in greater detail in Terziev *et al.* (2004). The development of the ontology continues under a collaborative 'community process' organized in accordance with the DILIGENT methodology, which is described in Chapter 9. In the following subsections, we provide an overview of its core module, its structure and some parts and design patterns more relevant to KM applications.

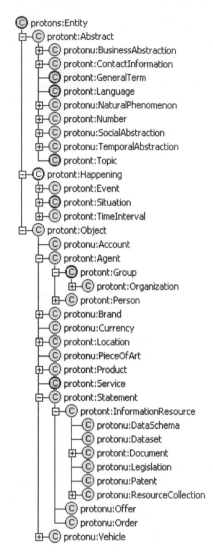

Figure 7.3   A view of the top part of the PROTON class hierarchy.

7.6.3. Scope, Coverage, Compliance

The extent of specialization of the ontology is partly determined on the basis of case studies within the scope of the SEKT project[17] and on a survey of the entity types in a corpus of general news (including political,

---

[17]http://www.sekt-project.com/

sports, and financial ones). The distribution of the most commonly used entity types varies greatly across domains. Still, as reported in Maynard *et al.* (2003), there are several general entity types that appear in the large majority of corpora (text collections) – `Person, Location, Organization, Money (Amount), Date`, etc. The proper representation and positioning of those basic types was one of the objectives in the design of PROTON and this was accomplished, for the most part, at the level of PROTON Top module layer.

The rationale behind PROTON is to provide a minimal, but nevertheless sufficient ontology, suitable for semantic annotation, as well as a conceptual basis for more general KM applications. Its predecessor— KIMO—was designed from scratch for use in the KIM system (http:// www.ontotext.com/kim/), which is described in Chapters 3 and X; a number of upper-level resources inspired its creation and development: OpenCyc (http://www.opencyc.org), Wordnet (http://www.cogsci. princeton.edu/~wn/), DOLCE (http://www.loa-cnr.it/DOLCE.html), EuroWordnet Top (Peters, 1998), and others.

One of the objectives in the development of PROTON has been to make it compliant with Dublin Core, the ACE annotation types,[18] and the ADL Feature Type Thesaurus.[19] This means that although these are not directly imported (for consistency reasons), a formal mapping of the appropriate classes and primitives is straightforward, on the basis of (i) compliant design and (ii) formal notes in the PROTON glosses, which indicate the appropriate mappings. For instance, in PROTON, a `hasContributor` property is defined, with a domain `Information-Resource` and a range `Agent`, as an equivalent of the `dc:contributor` element in Dublin Core. The development philosophy of PROTON is to make it compliant, in the future, with other popular standards and ontologies, such as FOAF.[20]

---

[18]The Automatic Content Extraction (ACE) is one of the most influential Information Extraction programs, see http://www.itl.nist.gov/iad/894.01/tests/ace/. A set of entity types is defined within 'The ACE 2003 Evaluation Plan' (ftp://jaguar.ncsl.nist.gov/ace/ doc/ace_evalplan-2003.v1.pdf). These are: Person, Organization, GPE (a Geo-Political Entity), Location, Facility.

[19]Alexandria Digital Library (ADL) is a project at the University of California, Santa Barbara, http://www.alexandria.ucsb.edu/. The Feature Type Thesaurus (FTT) can be found at. http://www.alexandria.ucsb.edu/gazetteer/FeatureTypes/ver070302/ index.htm. The `Location` branch of PROTON contains about 80 classes aligned with the FTT, which in its turn is aligned with the geographic feature designators of the GNS database of National Imagery and Mapping Agency of United States, (NIMA) at http:// earth-info.nga.mil/gns/html/. More details on the alignment are provided in Manov *et al.* (2003).

[20]The *Friend of a Friend* (FOAF) project is about creating a Web of machine-readable homepages describing people, the links between them and the things they create and do. See http://www.foaf-project.org/

7.6.4. The Architecture of Proton

PROTON is organized in three levels, including four modules. In Figure 7.4, the levels are layered from left to right. The System ontology module occupies the first, basic layer; then the Top, and Upper, and KM ontology modules are provided on top of it to form the diacritical modular architecture of PROTON.

The System module is an application ontology, which defines several notions and concepts of a technical nature that are substantial for the operation of any ontology-based software, such as semantic annotation and knowledge access tools. It includes the class `protons:Entity`—the top ('master') class for any sort of real-world objects and things, which could be of interest in some areas of discourse. In the system ontology it is defined that entities (i.e., the instances of `protons:Entity`) could have multiple names (instances of `protons:Alias`), that information about them could be extracted from particular `protons:EntitySource`-s, etc.

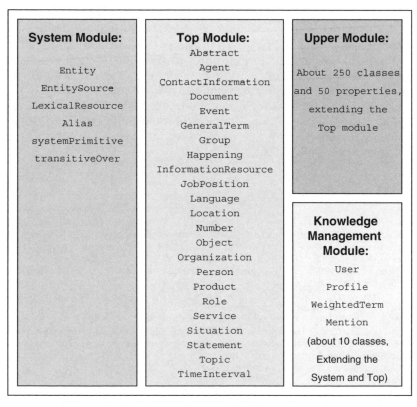

Figure 7.4   PROTON (PROTo ONtology) modules.

The Top ontology module starts with some basic philosophically reasoned distinctions between entity types, such as `Object`—existing entities, such as agents, locations, vehicles; `Happening`—events and situations; `Abstract`—abstractions that are neither objects nor happenings. The design at the highest level of the Top module follows the stratification principles of DOLCE, through the establishment of the PROTON trichotomy of Objects (`dolce:Endurant`), Happenings (`dolce:Perdurant`), and Abstracts (`dolce:Abstract`). The same stratification is also defined in Peters (1998). According to many experts in upper-level ontology construction (Guarino, 1998a; Peters, 1998), an important ontology design principle is that the extensions of these three branches should be disjoint, that is no individual should be an instance of more than one of these three top classes. One of the reasons for the introduction of this guiding principle is to avoid the 'overloading' of the subsumption (sub-class-of, is-a) relation.

These three classes are further specialized by about 20 general classes. These include `Agent`, `Person`, `Organization`, `Location`, `Event`, `InformationResource`, besides abstract notions, such as `Number`, `TimeInterval`, `Topic` (see the subsection below), and `GeneralTerm`. The featured entity types have their characteristic attributes and relations defined for them (e.g., `subRegionOf` property for `Location`-s, `hasPosition` for `Person`-s; `locatedIn` for `Organization`-s, `hasMember` for `Group`-s, etc.).

PROTON extends into its third layer, where two independent ontologies, which define much more specific classes, can be used: the PROTON Upper module and the PROTON KM module. Examples from the Upper module are: `Mountain`, as a specific type of `Location`; `ResourceCollection` as a sub-class of `InformationResource`. Having this ontology as a basis, one could easily add domain-specific extensions.

7.6.5. Topics in Proton

Based on the arguments, provided in the section on Topic-ontologies above, the following principles were adopted in the PROTON implementation:

- The class hierarchy of the schema ontology should not be mixed with topic hierarchies. One additional argument for this is that the latter can be expected to be specific for the different domains and applications. A further technical argument is that representing topics as instances of the Topic class avoids the computational intractability inherent in allowing classes as property values.
- We should avoid extensive modeling of semantics of topics using extensional semantics, as discussed earlier.

The `Topic` class (within the PROTON Top module) is meant to serve as a bridge between topic- and schema-ontologies. The specific topics should be defined as instances of the `Topic` class (or of a sub-class of it). The topic hierarchy is built using the `subTopic` property as a specialized subsumption relation between the topics. The latter is defined to be transitive but, importantly, it is not related to the `rdfs:subClassOf` meta-property. Typically, the instances of `Topic` are used as values of the `hasSubject` property (equivalent to `dc:subject`) of the `InformationResource` class.

`Topic` is any sort of a topic or a theme, explicitly defined for classification purposes. While any other class or entity could play the role of a topic in principle, the instances of class `Topic` are the only concepts in PROTON which are defined to serve as topics.[21] The `Topic` class is the natural top-class for linkage of logically informal taxonomies.

PROTON does not provide any `Topic` sub-classes as part of its Upper module layer. However, `Topic` is in certain relations with some of the classes in the KM module: `Profile` is related to `Topic` through property `isInterestedIn`; `Topic` is relater to `WeightedTerm` through property `hasWeightedTerm`.

An example for modeling of topics is given in Figure 7.5. Suppose one needs to encode that a particular document is about Jazz, using the

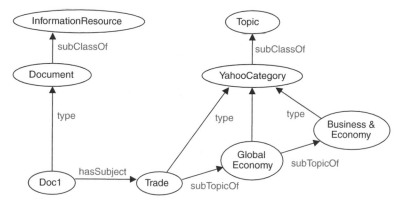

Figure 7.5 Topic modeling example: classifying a document by YahooCategory.

---

[21]For instance, the PROTON class `PublicCompany` can be intuitively used as a topic (e.g., 'documents about public companies'). PROTON suggests that this class should not be used as topic; instead, `PublicComapaniesTopic` should be defined as an instance of the `Topic` Class. It is often useful to link intuitively related concepts (as the two ones about public companies in the preceding example)—there is currently no support for such linking in PROTON. Such can however be added through an OWL annotation property named, for instance, `hasRelatedTopic`. Annotation properties are the only safe way of introducing properties relating classes and instances without escalating the complexity of the ontology to OWL Full.

Yahoo!® category hierarchy. *Jazz, Genre*, and *Music* are all instances of *YahooCategory*, which is a sub-class of `Topic`.

7.6.6. PROTON Knowledge Management Module

The KM module is in a sense an application-specific extension of PROTON, which introduces some definitions necessary for KM applications. The KM module is dependent on the System and Top modules. A snapshot from the KM module is given in Figure 7.6.

The remainder of this section describes the most important classes in the KM module.

7.6.6.1. Information Space

'Information spaces' denote collections of themed information resources (e.g., documents, maps, etc.). For example, the information space 'e-commerce' contains collections of documents relating to activities and entities concerning electronic commerce. The `InformationSpace` class is a specialization of `Agent`, and can be described as denoting a set of `User`'s personalized set of information 'items' in a specific milieu (e.g., a digital library or an online shopping portal). Each `Information-Space` is linked to an `InformationSpaceProfile` by means of the property `hasISprofile`, thus effectively modeling an `Information-Space` as a set of `Topics` (see later discussion on profiling).

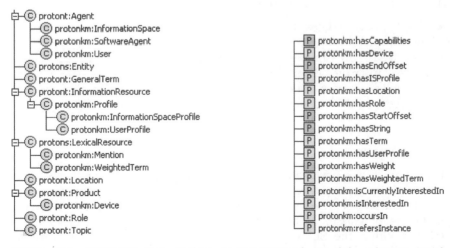

Figure 7.6 PROTON knowledge management module classes and properties.

7.6.6.2. Software Agent

SoftwareAgent is a specialization of Agent and denotes an artificial agent, which operates in a software environment. No proprietary properties are associated to this class.

7.6.6.3. User

The concept of a user is central for knowledge management, since a key aim is to represent a user's interests and context so that personalized, timely, relevant knowledge is provided. User is a specialization of Agent and designates a human user, who plays a Role with respect to some system. Every User has a UserProfile (related via the property hasUserProfile) and this is how the relation between a User and the Role he/she plays is realized. Each User can have several UserProfile-s, depending on his/her location, device, etc. This means that the hasUserProfile relation is a one-to-many relation.

7.6.6.4. Profile

Every User has a profile, and every InformationSpace has a profile associated with it. The class Profile is a subclass of InformationResource. It has two specializations: InformationSpaceProfile and UserProfile. Profiles can be linked to instances of Topic through the isInterestedIn property.

7.6.6.5. User Profile

The properties of a UserProfile are defined as follows:

- hasDevice—relates UserProfile with the Device class, representing the current device to which the user has access.
- hasLocation—relates UserProfile with the Location class, representing the current location of the user.
- hasRole—a user may have one or more roles which they switch between, so this relation links the UserProfile with a Role.

7.6.6.6. Mention

Mention is a specialization of LexicalResource (from the System module). Its main purpose is to model annotations (roughly speaking, identification of text strings in documents—e.g., 'London'—with instances or classes in the ontology—e.g, a specific instance of class City. Within the SEKT portfolio, for example, there is software to create annotations from a Document or an InformationResource. In this context, a Mention represents the mention of an Entity or a class in an `InformationResource`. The proprietary properties of `Mention` are:

- `hasStartOffset`—start offset in the content of the information resource;
- `hasEndOffset`—end offset in the content of the information resource;
- `hasString`—the string of the annotation, if such;
- `occursIn`—relates `Mention` with `InformationResource`;
- `refersInstance`—relates `Mention` with `Entity`.

7.6.6.7. Weighted Term

`WeightedTerm` is a sub-class of `LexicalResource`. It is closely connected to `Topic`—each `Topic` instance may have several `WeightedTerm`-s assigned to it via the `hasWeightedTerm` property. The `hasWeightedTerm` relation is a one-to-many relation, that is each `WeightedTerm` instance is associated with at most one `Topic` instance. A `GeneralTerm` can be related to multiple `Topic`-s and vice versa. Formally, `WeightedTerm` is related to `GeneralTerm` through property `hasTerm`. Weighted term is *de facto* an auxiliary class, through which a ternary predicate, the 'weighted' relation between a term and a topic can be modeled. Property `hasWeight` provides a relation between `WeightedTerm` and a real number that expresses the 'weight' of the term.

7.6.6.8. Device

The `Device` class is a specialization of `Product` (from the Top module). A `User` can use one or more `Device`-s for his/her activities regarding information resource search, management, usage, etc. This relation can be realized via the property `hasDevice` (proprietary to the `UserProfile` class), which relates the user profile of the user with the device(s) this user works with. Another property of `Device` is the `hasCapabilities` relation, which is designed to provide a relation between `Device` and a new 'Capability' class. Chapter 8 describes the use of the CC/PP device profiling ontology (linked to PROTON) to represent and reason about device properties in order to deliver information in a form suitable for a given device.

7.7. CONCLUSION

This chapter has presented an account of the use of ontologies in the KM context: what are the benefits; what sorts of data can be distinguished from the semantic and structural points of view and what is the relation between ontologies and data; what types of ontologies can be distinguished and for which task is each type appropriate. To provide a possible design for a basic ontology for KM and Semantic Web applications, we presented the PROTON ontology; it has proven to serve well as a database-schema replacement as well as a framework for semantic annotation; see (Kiryakov *et al.*, 2005). The usability of the ontology in KM applications is currently being tested in the various tools and case studies of the SEKT project, as discussed in Chapters 11 and 12. PROTON is being further developed under a community process organized in accordance with the DILIGENT methodology, described in Chapter 9.

REFERENCES

Beckett D. 2004. *RDF/XML Syntax Specification (Revised)*. http://www.w3.org/TR/2004/REC-rdf-syntax-grammar-20040210/

Borst P, Akkermans H, Top J. 1997. *Engineering ontologies. International Journal of Human-Computer Studies* 46:365–406.

Brickley D, Guha RV (eds). 2000. *Resource Description Framework (RDF) Schemas*, W3C http://www.w3.org/TR/2000/CR-rdf-schema-20000327/

Chinchor N, Robinson P. 1998. *MUC-7 Named Entity Task Definition* (version 3.5). In *Proceeding of the MUC-7*.

Davies J, Boncheva K, Manov D. 2004. *D5.0.1 Ontology Engineering in SEKT* (*internal project report*).

DCMI Usage Board. 2003b. *DCMI Type Vocabulary*. http://dublincore.org/documents/2003/11/19/dcmi-type-vocabulary/

DCMI Usage Board. 2005. *DCMI Metadata Terms*. http://dublincore.org/documents/2005/06/13/dcmi-terms/

Dean M, Schreiber G (eds), Bechhofer S, van Harmelen F, Hendler J, Horrocks I, McGuinness DL, Patel-Schneider PF, Stein LA. 2004. *OWL Web Ontology Language Reference*. W3C Recommendation February 10, 2004. http://www.w3.org/TR/owl-ref/

Fowler M. 2003. *UML Distilled: A Brief Guide to the Standard Object Modeling Language* (3rd ed.). Addison-Wesley.

Genesereth MR, Fikes R (eds). 1998. *Knowledge Interchange Format draft proposed American National Standard (dpANS)*. NCITS.T2/98-004. http://logic.stanford.edu/kif/

Gruber TR. 1992. *A translation approach to portable ontologies*. Knowledge Acquisition **5**(2):199–220, 1993. http://ksl-web.stanford.edu/KSL_Abstracts/KSL-92-71.html

Gruber TR. 1993. *Toward principles for the design of ontologies used for knowledge sharing*. In Guarino N, Poli R (eds). *International Workshop on Formal Ontology*, Padova, Italy, 1993. http://ksl-web.stanford.edu/KSL_Abstracts/KSL-93-04.html

Guarino N. 1998a. *Some Ontological Principles for Designing Upper Level Lexical Resources*. In Rubio A, Gallardo N, Castro R, Tejada A (eds), *Proceedings of First International Conference on Language Resources and Evaluation*. ELRA—European Language Resources Association, Granada, Spain, May 28–30, 1998, pp 527–534.

Guarino N, 1998b. *Formal Ontology in Information Systems*. In Guarino N (ed.), Formal Ontology in Information Systems. *Proceedings of FOIS'98*, Trento, Italy, June 6–8, 1998. IOS Press: Amsterdam, pp 3–15.

Guarino N, Giaretta P. 1995. *Ontologies and knowledge bases: Towards a terminological clarification*. In *Towards Very Large Knowledge Bases: Knowledge Building and Knowledge Sharing*, Mars N (ed). IOS Press: Amsterdam. pp 25–32.

Mahesh K, Nirenburg S, Cowie J, Farwell D. *An Assessment of Cyc for Natural Language Processing*. MCCS Report, New Mexico State University, 1996.

Manov D, Kiryakov A, Popov B, Bontcheva K, Maynard D, Cunningham H. 2003. *Experiments with geographic knowledge for information extraction*. NAACL-HLT 2003, Canada. Workshop on the Analysis of Geographic References, May 31 2003, Edmonton, Alberta.

Martin-Recuerda F, Harth A, Decker S, Zhdanova A, Ding Y, Stollberg M 2004. *Deliverable D2.1 'Report on requirements analysis and state-of-the-art'* within WP2 'Ontology Management' of the DIP project. https://bscw.dip.deri.ie/bscw/bscw.cgi/0/3012

Maynard D, Tablan V, Bontcheva K, Cunningham H, Wilks Y. 2003. *Multi-Source Entity recognition—An Information Extraction System for Diverse Text Types*. Technical report CS–02–03, University of Sheffield, Department of CS, 2003. http://gate.ac.uk/gate/doc/papers.html

Meyer-Fujara J, Heller B, Schlegelmilch S, Wachsmuth I. 1994. *Knowledge-level modularization of a complex knowledge base*. In *KI-94: Advances in Artificial Intelligence* Nebel B, Dreschler-Fischer L (eds). Springer: Berlin. pp 214–225.

Kiryakov A, Popov B, Ognyanov D, Manov D, Kirilov A, Goranov M. 2004a. *Semantic Annotation, Indexing, and Retrieval*. To appear in Elsevier's Journal of Web Semantics, Vol. 1, ISWC2003 special issue (2), 2004. http://www.websemanticsjournal.org/

Kiryakov A, Ognyanov D, Kirov V. 2004b. *D2.2: An Ontology Representation and Data Integration (ORDI) Framework*. DIP project deliverable. http://dip.semanticweb.org

Kiryakov A, Ognyanov D, Manov D. 2005. *OWLIM—A Pragmatic Semantic Repository for OWL*. In *Proceeding of International Workshop on Scalable Semantic Web Knowledge Base Systems* (SSWS 2005), WISE 2005, 20 November, New York City, USA.

Klyne G, Carroll JJ. 2004. *Resource Description Framework (RDF): Concepts and Abstract Syntax*. W3C recommendation 10 February, 2004. http://www.w3.org/TR/rdf-concepts/

Laboratory of Applied Ontologies, Institute of Cognitive Science and Technology, Italian National Research Council. *DOLCE: A Descriptive Ontology for Linguistic and Cognitive Engineering*. http://www.loa-cnr.it/DOLCE.html

Noy N 2004 *Representing Classes As Property Values on the Semantic Web*. W3C Working Draft 21 July 2004. http://www.w3.org/TR/2004/WD-swbp-classes-as-values-20040721/

Peters W (ed.). 1998. *The EuroWordNet Base Concepts and Top Ontology*. Version 2, Final. January 22, 1998. http://www.illc.uva.nl/EuroWordNet/corebcs/topont.html

Pinto S, Staab S, Tempich C. 2004. *DILIGENT: Towards a fine-grained methodology for Distributed Loosely-controlled and evolvInG Engineering of oNTologies*. In Proceedings of ECAI-2004, Valencia, August 2004.

Shirky C. 2005. *Ontology is Overrated: Categories, Links, and Tags*. Clay Shirky's Writings About the Internet. Economics & Culture, Media & Community. http://www.shirky.com/writings/ontology_overrated.html

Spark Jones K. 2004. *What's new about the Semantic Web? Some questions*. SIGIR Forum December 2004, Volume 38 Number 2. http://www.sigir.org/forum/2004D-TOC.html

Terziev I, Kiryakov A, Manov D. 2004. *D 1.8.1. Base upper-level ontology (BULO) Guidance,* report EU-IST Integrated Project (IP) IST-2003-506826 SEKT), 2004. http://proton.semanticweb.org/D1_8_1.pdf

Williams S. 2002. *The Associative Model of Data.* Second Edition, Lazy Software, Ltd. ISBN: 1-903453-01-1. http://www.lazysoft.com

8

Semantic Information Access

Kalina Bontcheva, John Davies, Alistair Duke, Tim Glover, Nick Kings
and Ian Thurlow

8.1. INTRODUCTION

Previous chapters have described the core technologies which underpin
the Semantic Web. This chapter describes how these semantic web
technologies can provide an improved user experience, through
enhanced tools for accessing knowledge. The domain model implicit
in an ontology can be used as a unifying structure to give information
about a common representation and semantics. Once this unifying
structure for heterogeneous information sources exists, it can be
exploited to improve the performance of knowledge access tools. In
this chapter, we look at the application of such technology to three
aspects of knowledge access: semantic search and browse tools; the
generation of information expressed in natural language from formal
(ontological) knowledge bases (natural language generation); and the
intelligent delivery of information to multiple end-user information
appliances (device independence). Finally, we describe SEKTAgent, a
knowledge management tool which illustrates the use of all three of
these technologies.

8.2. KNOWLEDGE ACCESS AND THE SEMANTIC WEB

We begin this section by discussing the shortcomings of current search
technology, before looking at how semantic web technology can offer the
user a better search experience, and describing a number of systems

Semantic Web Technologies: Trends and Research in Ontology-based Systems
John Davies, Rudi Studer, Paul Warren © 2006 John Wiley & Sons, Ltd

which have been developed to search semantically annotated information resources.

8.2.1. Limitations of Current Search Technology

8.2.1.1. Query Construction

In general, when specifying a search, users enter a small number of terms in the query. Yet the query describes the information need, and is commonly based on the words that people expect to occur in the types of document they seek. This gives rise to a fundamental problem, in that not all documents will use the same words to refer to the same concept. Therefore, not all the documents that discuss the concept will be retrieved by a simple keyword-based search. Furthermore, query terms may of course have multiple meanings (query term *polysemy*). As conventional search engines cannot interpret the sense of the user's search, the ambiguity of the query leads to the retrieval of irrelevant information.

Although the problems of query ambiguity can be overcome to some degree, for example by careful choice of additional query terms, there is evidence to suggest that many people may not be prepared to do this. For example, an analysis of the transaction logs of the Excite WWW search engine (Jansen *et al.*, 2000) showed that web search engine queries contain on average 2.2 terms. Comparable user behaviour can also be observed on corporate Intranets. An analysis of the queries submitted to BT's Intranet search engine over a 4-month period between January 2004 and May 2004 showed that 99 % of the submitted queries only contained a single phrase and that, on average, each phrase contained 1.82 keywords.

8.2.1.2. Lack of Semantics

Converse to the problem of polysemy, is the fact that conventional search engines that match query terms against a keyword-based index will fail to match relevant information when the keywords used in the query are different from those used in the index, despite having the same meaning (index term *synonymy*). Although this problem can be overcome to some extent through thesaurus-based expansion of the query, the resultant increased level of document recall may result in the search engine returning too many results for the user to be able to process realistically.

In addition to an inability to handle synonymy and polysemy, conventional search engines are unaware of any other semantic links between concepts. Consider, for example, the following query:

'telecom company' Europe 'John Smith' director

The user might require, for example, documents concerning a telecom company in Europe, a person called John Smith, and a board appointment. Note, however, that a document containing the following sentence would not be returned using conventional search techniques:

'At its meeting on the 10th of May, the board of London-based O2 appointed John Smith as CTO'

In order to be able to return this document, the search engine would need to be aware of the following semantic relations:

'O2 is a mobile operator, which is a kind of telecom company;
London is located in the UK, which is a part of Europe;
A CTO is a kind of director.'

8.2.1.3. Lack of Context

Many search engines fail to take into consideration aspects of the user's context to help disambiguate their queries. User context would include information such as a person's role, department, experience, interests, project work etc. A simple search on BT's Intranet demonstrates this. A person working in a particular BT line of business searching for information on their corporate clothing entitlement is presented with numerous irrelevant results if they simply enter the query 'corporate clothing'. More relevant results are only returned should the user modify their query to include further search terms to indicate the part of the business in which they work. As discussed above, users are in general unwilling to do this.

8.2.1.4. Presentation of Results

The results returned from a conventional search engine are usually presented to the user as a simple ranked list. The sheer number of results returned from a basic keyword search means that results navigation can be difficult and time consuming. Generally, the user has to make a decision on whether to view the target page based upon information contained in a brief result fragment. A survey of user behaviour on BT's intranet suggests that most users will not view beyond the 10th result in a list of retrieved documents. Only 17 % of searches resulted in a user viewing more than the first page of results.[1]

---

[1]Out of a total of 143 726 queries submitted to the search engine, there were 251 192 occasions where a user clicked to view more than the first page of results. Ten results per page are returned by default.

8.2.1.5. Managing Heterogeneity

Corporate search engines are required to index a wide range of subject material from a diverse and distributed collection of information sources, including web sites, content management systems, document management systems, databases and perhaps certain relevant areas of the external web. This represents a challenge not only in simple terms of connectivity to multiple information resources, but also in providing a coherent view of diverse sources and types of information.

8.2.2. Role of Semantic Technology

Semantic technology has the potential to offer solutions to many of the limitations described above, by providing enhanced knowledge access based on the exploitation of machine-processable metadata. Central to the vision of the Semantic Web are ontologies. These facilitate knowledge sharing and reuse between agents, be they human or artificial. They offer this capability by providing a consensual and formal conceptualisation of a given domain. Information can then be annotated with respect to an ontology. This leads to distributed, heterogeneous information sources being unified through a machine-processable common domain model (ontology). Ontologies are populated with semantic metadata as discussed in more detail in Chapter 3. The PROTON ontology itself is introduced and discussed in Chapter 7.

Search engines based on conventional information retrieval techniques alone tend to offer high recall but lower precision. The user is faced with too many results and many results that are irrelevant due to a failure to handle polysemy and synonymy, still less any richer semantic relations.

As we will exemplify later in this chapter, the use of ontologies and associated metadata can allow the user to more precisely express their queries thus avoiding the problems identified above. Users can choose ontological concepts to define their query or select from a set of returned concepts following a search in order to refine their query. They can specify queries over the metadata and indeed combine these with full text queries if desired.

Furthermore, the use of semantic web technology offers the prospect of a more fundamental change to knowledge access. Current technology supports a process wherein the user attempts to frame an information need by specifying a query in the form of either a set of keywords or a piece of natural language text. Having submitted a query, the user is then presented with a ranked list of documents of relevance to the query. However, this is only a partial response to the user's actual requirement which is for *information* rather than lists of documents.

It is suggested here, therefore, that the future of search engines lies in supporting more of the information management process, as opposed to seeking incremental and modest improvements to relevance ranking of documents. In this approach, software supports more of the process of analysing relevant documents rather than merely listing them and leaving the rest of the information analysis task to the user. Corporate knowledge workers need information defined by its meaning, not by text strings ('bags of words'). They also need information relevant to their interests and to their current context. They need to find not just documents, but sections and information entities within documents and even digests of information created from multiple documents. As described below, the exploitation of metadata and ontological information can offer this information-centric approach, as opposed to the prevailing document-centric technology.

8.2.3. Searching XML

The eXtensible Mark-up Language (XML), a specification for machine-readable documents, is one of the first steps towards a Semantic Web. XML is a meta-language, as such, it provides a mechanism for representing other languages in a standardised way. XML mark-up describes (and prescribes) a document's data layout and structure as a tree of nested tags. XML-based search engines exploit this mark-up, enabling searches for documents where keywords and phrases appear within the elements of an XML document, for example search for the phrase 'Semantic Web' within all ⟨title⟩ elements of a set of XML documents.

In an early implementation of XML-aware search, QuizXML (Davies, 2000) compiles a list of the 'tags' that annotate and subdivide the documents within which document terms are found. QuizXML then creates a finer-grained index than traditional search engines: its index maps keywords to both the documents *and* the XML tags within which those keywords are found. QuizXML allows users to explore interactively the list of tags in which a given query occurs, and select a particular tag in order to refine the search results to only those documents where the search query occurs in a part of an XML document marked up by the selected tag.

In a more sophisticated approach, described in Cohen *et al.* (2003), the XSearch semantic search engine has been designed to return semantically-related document *fragments* in response to a user's query (in preference to returning a reference to the complete document). This is particularly useful in cases where a large document contains information in addition to that which matches the query, but which is not necessarily related to the query. XSearch provides a simple query interface that does not require a detailed knowledge of the structure of the XML documents being sought. The XSearch query syntax enables the user to specify how query terms must be related to the XML tags (but does not enforce this—indeed, a query containing only keywords may be entered).

XSearch incorporates techniques for determining which elements of an XML document are semantically related. An answer to an XSearch query typically contains multiple document fragments which are so related. Results ranking takes into account both keyword relevance and the degree of semantic relationship. Experiments reported in (Cohen *et al.*, 2003) indicate that XSearch is efficient and scalable.

In related work, the XRANK XML search engine (Guo *et al.*, 2003) also returns document fragments in response to queries. In XRANK, each query term is matched against both document content and document mark-up (XML tags). It ranks the elements of an XML document using an extension of the Google PageRank algorithm (Brin and Page, 1998), which (in part) uses the hyperlinks on a webpage to assess its relevance and importance. The ranking algorithm in XRANK is analogous but is performed at the level of the hyperlink structure of the document *elements* rather than at the level of the XML document as a whole. It then ranks the answers to a query by taking into account both the ranking for the elements and keyword proximity. Proximity in this context is not simply determined by position in the document but also by position in the XML tree structure (roughly speaking, measured by the distance to the lowest common ancestor in the tree).

Cohen *et al.* (2002) and Florescu *et al.* (2000) describe other examples of support for keyword querying over XML content.

Although XML promotes greater interoperability between applications that conform to a pre-defined data standard, the use and semantics of each XML tag are not defined. For example, the semantics implicit in the embedding of one tag within another are hidden. Interpretation of XML tags relies on the implicit knowledge that is hardcoded into the application programs that access the XML encoded information (or indeed by the person being presented with the search results from an XML enabled search engine).

8.2.4. Searching RDF

The resource description framework (RDF; W3C, 2004) is a W3C standard for data interchange. RDF builds on XML to provide a mechanism for describing data about resources on the web (e.g. documents) in terms of named properties and values. RDF provides a model that enables information about any resource to be encoded in a formal, machine-processable format. RDF data is serialisable in XML. An RDF description of a resource consists of a set of RDF statements (triples). Each statement has a uniform structure consisting of three parts: an object (a resource), an attribute (a property) and a value (another resource or plain text), for example a Web page (the resource) was 'created by' (the attribute) 'John Smith' (a value). RDF for example enables information to be expressed in a formal way that software agents can read, process and act upon. RDF

Schema (RDFS) extends RDF by providing a mechanism that enables RDF documents to convey intended meaning. It provides the semantic information required by an application to correctly interpret the RDF statements. RDFS defines an ontological vocabulary in the form of classes, class properties and relationships, for example descriptive terms such as *book, author, title, hasName* etc. RDF and RDFS (often denoted together as RDF(S)), when combined, provide the syntactic model and semantic structure for defining machine-processable ontologies. Thus, the use of RDF(S) to annotate content goes beyond what is possible in XML in a number of ways:

- RDF(S) is descriptive not prescriptive—XML dictates the format of individual documents; whereas RDF(S) allows the description of any content and the RDF(S) annotations need not be embedded within the content itself.
- More than one RDF(S) ontology can be used to describe the same content for different purposes if required.
- RDF(S) has a well-defined semantics regarding, for example the subclass relation.
- RDF(S) allows the definition of a set of relations between resources as described above.

The QuizRDF search engine (Davies *et al.*, 2003) combines free-text search with a capability to exploit RDF annotation querying. QuizRDF combines search and browsing capabilities into a single tool and allows RDF annotations to be exploited (searched over) where they exist, but will still function as a conventional search engine in the absence of those annotations.

The user enters a query into QuizRDF as they would in a conventional search engine. A list of documents is returned, ranked according to the resource's relevance to the user's query using a traditional vector space approach (Salton *et al.*, 1997). The QuizRDF data model assumes each annotated document is linked to one *or more* subjects via an 'isAbout' property.[2] The subjects of the documents returned in response to a query are ascertained and displayed along with the traditional ranked list.

By selecting one of the displayed subjects (ontological classes), the user can filter the retrieval list to include only those resources (documents) that are instances of the selected class. QuizRDF also displays the properties and classes related to the selected class. Each class displayed has an associated hyperlink that allows the user to browse the RDFS ontology: clicking on the class name refreshes the display to show that class properties and related classes[3] and again filters the results list to

---

[2]This corresponds closely to the PROTON hasSubject property described in Chapter 7.
[3]A related class in this context is a class which is the domain or range of a property of the original class.

show only URIs of the related class. Where properties have literal types
(e.g., string) as their range, QuizRDF enables users to query against these
properties. So for example the class Painting has a property hasTechni-
que with range of type 'string'. If a set of documents has been returned of
class Painting, a user could enter 'oil on canvas' as a value for this
property and the document list would be filtered to show only those
documents which have the value 'oil on canvas' for this property.

Thus QuizRDF has two retrieval channels: a keyword query against
the text and a much more focussed query against specific RDF properties,
as well as supporting ontology browsing. Searching the full-text of the
documents ensures the desired high recall in the initial stages of the
information-seeking process. In the later stages of the search, more
emphasis can be put on searching the RDF annotations (property values)
to improve the precision of the search.

Experiments with QuizRDF on an RDF-annotated web site showed
improvements in performance as compared to a conventional keyword-
based search facility (Iosif *et al.*, 2003).

8.2.5. Exploiting Domain-specific Knowledge

As discussed above, conventional search engines have no model of how
the concepts denoted by query terms may be linked semantically. When
searching for a paper published by a particular author, for example it
may be helpful to retrieve additional information that relates to that
author, such as other publications, curriculum vitae, contact details etc. A
number of search engines are now emerging that use techniques to apply
ontology-based domain-specific knowledge to the indexing, similarity
evaluation, results augmentation and query enrichment processes.

Rocha *et al.* (2004) describe a search architecture that applies a
combination of spread activation and conventional information retrieval
to a domain-specific semantic model in order to find concepts relevant to
a keyword-based query. The essence of spread activation, as applied in
conventional textual searching, is that a document may be returned by a
query, even if it contains none of the query keywords. This happens if the
document is linked to by many other documents which do contain the
keywords. The inference is that such a document will very likely be
relevant to the query, despite not possessing the keywords.

In the case described here, the user expresses his query as a set of
keywords. This query is forwarded to a conventional search engine
which assigns a score to each document in its index in the usual way.

In addition to a conventional index, the system contains a domain-
specific knowledge base. This knowledge base consists of a model of the
domain, with instance nodes pointing to web resources, each node
having additional data in the form of linked properties as specified in
the domain model (ontology). Weightings that express the strength of

each instance relation in the ontology are then derived. A number of different approaches to this derivation are taken and the authors state that it is not possible to find a weight derivation formula which is optimal for all application areas. Thus the resulting network has, for each relationship, a semantic label and an associated numerical weight. The intuition behind this approach is that better search results can be obtained by exploiting not only the relationships within the ontology, but also the strength of those relationships.

Searching proceeds in two phases: as mentioned, a traditional approach is first used to derive a set of documents from a keyword-based query. As discussed, these documents are associated with instances in the ontology and are linked via weighted relations to other nodes. This set of nodes is supplied as an initial set to the spread activation algorithm, using the numeric ranking from the traditional retrieval algorithm as the initial activation value for each node. The set of nodes obtained at the end of the propagation are then presented as the search results. Two case studies are reported, showing that in these cases the combination of the traditional and spread activation techniques performs better than either on its own.

Guha *et al.* (2003) describe a Semantic Web-based search engine (ABS—activity-based search) that aims to augment traditional search results when seeking information in relation to people, places, events, news items etc. The semantic search application, which runs as a client of the TAP infrastructure (Guha and McCool, 2003), sends a user-supplied query to a conventional search engine. Results returned from the conventional search engine are augmented with relevant information aggregated from distributed data sources that form a knowledge base (the information is extracted from relevant content on targeted web sites and stored as machine-readable RDF annotations). The information contained in the knowledge base is independent of and additional to the results returned from the conventional search engine. A search for a musician's name, for example would augment the list of matching results from the conventional search engine with information such as current tour dates, discography, biography etc. Figure 8.1 shows a typical search result from ABS.

Of course, the uptake of the Semantic Web (and semantic Intranet) will depend upon the availability of ontologies and metadata associated with Web (Intranet) content. Currently, this metadata is not available in abundance. The manual acquisition of large amounts of metadata is generally considered to be impractical. It is therefore necessary to acquire or generate the metadata, either automatically or at least semi-automatically.

In a further development on the idea of exploiting an ontological knowledge base to enhance search, the knowledge and information management (KIM) infrastructure (Popov *et al.*, 2003) aims to provide automated methods for semantic annotation as discussed in Chapter 3.

Figure 8.1 Semantic search with TAP.

The KIM architecture also facilitates the indexing and retrieval of documents with respect to particular named entities such as people, companies, organisations and locations. Formal background knowledge (held in the KIM ontological knowledgebase) can be linked to the entities identified in Web documents as described in Chapter 3. The knowledge base can host two types of entity knowledge: pre-populated descriptions and knowledge acquired from trusted sources, and automatically extracted descriptions derived through knowledge discovery and acquisition methods such as data mining. The KIM platform enables the ontology to be extended and knowledge base to be populated to meet the domain-specific needs of a semantic annotation application.

The semantic annotation process in KIM assigns to the named entities in the text links to semantic descriptions of those entities in the ontological knowledgebase. The semantic descriptions provide both class and instance information about the entities referred to in the documents. KIM analyses the text, recognises references to named entities and matches the reference with a known entity from the knowledgebase (i.e. an ontological instance). The reference in the document is annotated with the URI of the entity. In Chapter 3, Figure 3.2 shows the way in which a segment of text concerning a Bulgarian company might be associated with a number of entities in the ontological knowledge base.

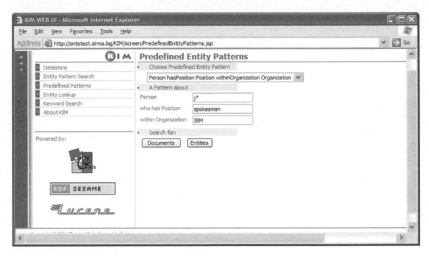

Figure 8.2 Semantic querying in KIM.

The semantic annotations can then be used for indexing and retrieval, categorisation, visualisation and smooth traversal between unstructured text and available relevant knowledge. The application of entity coreference resolution means that the system would regard the strings 'Tony Blair' 'Mr Blair' 'the Prime Minister' as referring to the same entity in the ontological knowledge base. Semantic querying is supported against the repository of semantically annotated documents. This would allow for example a query to be formulated that targets all documents that refer to *Persons* that hold a specified *Position* within an *Organisation*. Figure 8.2 shows a KIM ontological query concerning a person whose name begins with 'J', and who is a spokesman for IBM. Note that the user interface shown is for a specific query type (regarding people with specific positions in named organisations). A more general interface is also available, allowing the specification of queries about any type of entity, relations between such entities and required attribute values (e.g. 'find all documents referring to a `Person` that `hasPosition` "CEO" within a `Company`, `locatedIn` a `Country` with name "UK"'). To answer the query, KIM applies the semantic restrictions over the entities in the knowledge base. The resulting set of entities is matched against the semantic index and the referring documents are retrieved with relevance ranking according to these named entities.

Figure 8.3 Shows that four such ontological entities have been found in the documents indexed.

It is then possible to browse a list of documents containing the specified entities and KIM renders the documents, with entities from the query highlighted (in this example IBM and the identified spokesperson).

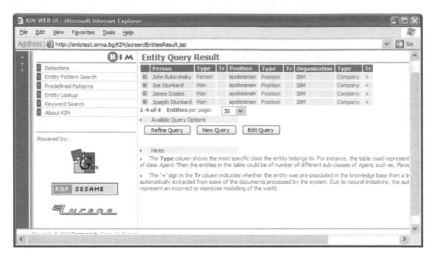

Figure 8.3 Semantic query results.

It should be noted that the work surveyed here is not claimed to be comprehensive, but indicative of the research being carried out in a large number of groups worldwide.

In other work, Berstein *et al.* (2005) describe a controlled language approach whereby a subset of English is entered by the user as a query and is then mapped into a semantic query via a discourse representation structure. Vallet *et al.* (2005) propose an ontology-based information retrieval model using a semantic indexing scheme based on annotation weighting techniques.

8.2.6. Searching for Semantic Web Resources

We have seen in the earlier sections a variety of approaches for searching semantically annotated information resources. The Swoogle search engine (Ding *et al.*, 2004) is tackling a related but different problem: it is primarily concerned with finding ontologies and related instance data.

Finding ontologies is seen as important to avoid the creation of new ontologies where serviceable ones already exist. It is hoped that this approach will lead to the emergence of widely-used canonical ontologies. Swoogle supports querying for ontologies containing specified terms. This can be refined to find ontologies where such terms occur as classes or properties, or to find ontologies that are in some sense about the specified term (as determined by Swoogle's ontology retrieval engine). The ontologies thus found are ranked according to Swoogle's OntologyRank algorithm which attempts to measure the degree to which a given ontology is used.

In order to offer such search facilities, Swoogle builds an index of semantic web documents (defined as web-accessible documents written in a semantic web language). A specialised crawler has been built using a range of heuristics to identify and index semantic web documents.

The creators of Swoogle are building an ontology dictionary based on the ontologies discovered by Swoogle.

8.2.7. Semantic Browsing

Web browsing complements searching as an important aspect of information-seeking behaviour. Browsing can be enhanced by the exploitation of semantic annotations and below we describe three systems which offer a semantic approach to information browsing.

Magpie (Domingue *et al.*, 2004) is an internet browser plug-in which assists users in the analysis of web pages. Magpie adds an ontology-based semantic layer onto web pages on-the-fly as they are browsed. The system automatically highlights key items of interest, and for each highlighted term it provides a set of 'services' (e.g. contact details, current projects, related people) when you right click on the item. This relies, of course, on the availability of a domain ontology appropriate to the page being browsed.

CS AKTiveSpace (Glaser *et al.*, 2004) is a semantic web application which provides a way to browse information about the UK Computer Science Research domain, by exploiting information from a variety of sources including funding agencies and individual researchers. The application exploits a wide range of semantically heterogeneous and distributed content. AKTiveSpace retrieves information related to almost two thousand active Computer Science researchers and over 24 000 research projects, with information being contained within 1000 published papers, located in different university web sites. This content is gathered on a continuous basis using a variety of methods including harvesting publicly available data from institutional web sites, bulk translation from existing databases, as well as other data sources. The content is mediated through an ontology and stored as RDF triples; the indexed information comprises around 10 million RDF triples in total.

CS AKTive Space supports the exploration of patterns and implications inherent in the content using a variety of visualisations and multi-dimensional representations to give unified access to information gathered from a range of heterogeneous sources.

Quan and Karger (2004) describe Haystack, a browser for semantic web information. The system aggregates and visualises RDF metadata from multiple arbitrary locations. In this respect, it differs from the two semantic browsing systems described above which are focussed on using metadata annotations to enhance the browsing and display of the data itself.

Presentations styles in Haystack are themselves described in RDF and can be issued by the content server or by context-specific applications which may wish to present the information in a specific way appropriate to the application at hand. Data from multiple sites and particular presentation styles can be combined by Haystack on the client-side to form customised access to information from multiple sources. The authors demonstrate a Haystack application in the domain of bioinformatics.

In other work (Karger *et al.*, 2003), it is reported that Haystack also incorporates the ability to generate RDF data using a set of metadata extractors from a variety of other formats, including documents in various formats, email, Bibtex files, LDAP data, RSS feeds, instant messages and so on. In this way, Haystack has been used to produce a unified Personal Information Manager. The goal is to eliminate the partitioning which has resulted from having information scattered between e-mail client(s), filesystem, calendar, address book(s), the Web and other custom repositories.

8.3. NATURAL LANGUAGE GENERATION FROM ONTOLOGIES

Natural Language Generation (NLG) takes structured data in a knowledge base as input and produces natural language text, tailored to the pre-sentational context and the target reader (Reiter and Dale, 2000). NLG techniques use and build models of the context, and the user and use them to select appropriate presentation strategies, for example to deliver short summaries to the user's WAP phone or a longer multi-modal text to the user's desktop PC.

In the context of the semantic web and knowledge management, NLG is required to provide automated documentation of ontologies and knowledge bases. Unlike human-written texts, an automatic approach will constantly keep the text up-to-date which is vitally important in the semantic web context where knowledge is dynamic and is updated frequently. The NLG approach also allows generation in multiple languages without the need for human or automatic translation (see (Aguado *et al.*, 1998)).

Generation of natural language text from ontologies is an important problem. Firstly, because textual documentation is more readable than the corresponding formal notations and thus helps users who are not knowledge engineers to understand and use ontologies. Secondly, a number of applications have now started using ontologies for knowledge representation, but this formal knowledge needs to be expressed in natural language in order to produce reports, letters etc. In other words, NLG can be used to present structured information in a user-friendly way.

There are several advantages to using NLG rather than using fixed templates where the query results are filled in:

- NLG can use different sentence structures depending on the number of query results, for example conjunction versus itemised list.
- Depending on the user's profile of their interests, NLG can include different types of information—affiliations, email addresses, publication lists, indications on collaborations (derived from project information).
- Given the variety of information which can be included and how it can be presented, and depending on its type and amount, writing templates may not be feasible because of the number of combinations to be covered. This variation in presentational formats comes from the fact that each user of the system has a profile comprising user supplied (or system derived) personal information (name, contact details, experience, projects worked on), plus information derived semi-automatically from the user's interaction with other applications. Therefore, there will be a need to tailor the generated presentations according to user's profile.

8.3.1. Generation from Taxonomies

PEBA is an intelligent online encyclopaedia which generates descriptions and comparisons of animals (Dale *et al.*, 1998). In order to determine the structure of the generated texts, the system uses text patterns which are appropriate for the fairly invariant structure of the animal descriptions. PEBA has a taxonomic knowledge base which is directly reflected in the generated hypertext because it includes links to the super- and sub-concepts (see example below). Based on the discourse history, that is what was seen already, the system modifies the page opening to take this into account. For example, if the user has followed a link to marsupial from a node about the kangaroo, then the new text will be adapted to be more coherent in the context of the previous page:

> *'Apart from the Kangaroo, the class of Marsupials also contains the following subtypes...' (Dale et al., 1998)*

The main focus in PEBA is on the generation of comparisons which improve the user's understanding of the domain by comparing the currently explained animal to animals already familiar to the user (from common knowledge or previous interaction).

The system also does a limited amount of tailoring of the comparisons, based on a set of hard-coded user models derived from stereotypes, for example novice or expert. These stereotypes are used for variations in language and content. For example, when choosing a target for a comparison, the system might pick cats for novice users, as they are commonly known animals.

8.3.2. Generation of Interactive Information Sheets

Buchanan *et al.* (1995) developed a language generator for producing concept definitions in natural language from the Loom knowledge representation language.[4] Similar to the ONTOGENERATION project (see below) this approach separates the domain model from the linguistic information. The system is oriented towards providing patients with interactive information sheets about illnesses (migraine in this case), which are tailored on the basis of the patient's history (symptoms, drugs etc). Further information can be obtained by clicking on mouse-sensitive parts of the text.

8.3.3. Ontology Verbalisers

Wilcock (2003) has developed general purpose ontology verbalisers for RDF and DAML + OIL (Wilcock *et al.*, 2003) and OWL. These are template based and use a pipeline of XSLT transformations in order to produce text. The text structure follows closely the ontology constructs, for example 'This is a description of John Smith identified by http:// ...His given name is John...' (Wilcock, 2003).

Text is produced by performing sentence aggregation to connect sentences with the same subject. Referring expressions like 'his' are used instead of repeating the person's name. The approach is a form of *shallow generation*, which is based on domain- and task-specific modules.

The language descriptions generated are probably more suitable for ontology developers, because they follow very closely the structures of the formal representation language, that is RDF or OWL.

The advantages of Wilcock's approach is that it is fully automatic and does not require a lexicon. In contrast, other approaches discussed here require more manual input (lexicons and domain schemas), but on the other hand they generate more fluent reports, oriented towards end users, not ontology builders.

8.3.4. Ontogeneration

The ONTOGENERATION project (Aguado *et al.*, 1998) explored the use of a linguistically oriented ontology (the Generalised Upper Model

---

[4]http://www.isi.edu/isd/LOOM/

(GUM) (Bateman *et al.*, 1995)) as an abstraction between language generators and their domain knowledge base (chemistry in this case). The GUM is a linguistic ontology with hundreds of concepts and relations, for example part-whole, spatio-temporal, cause-effect. The types of text that were generated are: concept definitions, classifications, examples and comparisons of chemical elements.

However, the size and complexity of GUM make customisation more difficult for nonexperts. On the other hand, the benefit from using GUM is that it encodes all linguistically-motivated structures away from the domain ontology and can act as a mapping structure in multi-lingual generation systems. In general, there is a trade-off between the number of linguistic constructs in the ontology and portability across domains and applications.

8.3.5. Ontosum and Miakt Summary Generators

Summary generation in ONTOSUM starts off by being given a set of RDF triples, for example derived from OWL statements. Since there is some repetition, these triples are first pre-processed to remove duplicates. In addition to triples that have the same property and arguments, the system also removes those triples with equivalent semantics to an already verbalised triple, expressed through an inverse property. The information about inverse properties is provided by the ontology (if supported by the representation formalism). An example summary is shown later in this chapter (Figure 8.6) where the use of ONTOSUM in a semantic search agent is described.

The lexicalisations of concepts and properties in the ontology can be specified by the ontology engineer, be taken to be the same as concept and property names themselves, or added manually as part of the customisation process. For instance, the AKT ontology[5] provides label statements for some of its concepts and instances, which are found and imported in the lexicon automatically. ONTOSUM is parameterised at run time by specifying which properties are to be used for building the lexicon.

A similar approach was first implemented in a domain- and ontology-specific way in the MIAKT system (Bontcheva *et al.*, 2004). In ONTOSUM it is extended towards portability and personalisation, that is lowering the cost of porting the generator from one ontology to another and generating summaries of a given length and format, dependent on the user target device.

Similar to the PEBA system, summary structuring is done using *discourse/text schemas* (Reiter and Dale, 2000), which are script-like

---

[5]http://www.aktors.org/ontology/

structures which represent discourse patterns. They can be applied recursively to generate coherent multi-sentential text. In more concrete terms, when given a set of statements about a given concept or instance, discourse schemas are used to impose an order on them, such that the resulting summary is coherent. For the purposes of our system, a coherent summary is a summary where similar statements are grouped together.

The schemas are independent of the concrete domain and rely only on a core set of four basic properties—active-action, passive-action, attribute, and part-whole. When a new ontology is connected to ONTOSUM, properties can be defined as a sub-property of one of these four generic ones and then ONTOSUM will be able to verbalise them without any modifications to the discourse schemas. However, if more specialised treatment of some properties is required, it is possible to enhance the schema library with new patterns, that apply only to a specific property.

Next ONTOSUM performs *semantic aggregation*, that is it joins RDF statements with the same property name and domain as one conceptual graph. Without this aggregation step, there will be three separate sentences instead of one bullet list (see Figure 8.5), resulting in a less coherent text.

Finally, ONTOSUM verbalises the statements using the HYLITE + sur-surface realiser, which determines the grammatical structure of the generated sentences. The output is a textual summary. Further details can be found in Bontcheva (2005).

An innovative aspect of ONTOSUM, in comparison to previous NLG systems for the Semantic Web, is that it implements tailoring and personalisation based on information from the user's device profile. Most specifically, methods were developed for generating summaries within a given length restriction (e.g., 160 characters for mobile phones) and in different formats – HTML for browsers and plain texts for emails and mobile phones (Bontcheva, 2005). The following section discusses a complementary approach to device independent knowledge access and future work will focus on combining the two.

Another novel feature of ONTOSUM is its use of *ontology mapping* rules, as described in Chapter 6 to enable users to run the system on new ontologies, without any customisation efforts.

8.4. DEVICE INDEPENDENCE: INFORMATION ANYWHERE

Knowledge workers are increasingly working both in multiple locations and while on the move using an ever wider variety of terminal devices. They need information delivered in a format appropriate to the device at hand.

The aim of device independence is to allow authors to produce content that can be viewed effectively, using a wide range of devices. Differences in device properties such as screen size, input capabilities, processing capacity, software functionality, presentation language and network protocols make it challenging to produce a single resource that can be presented effectively to the user on any device.

In this section, we review the key issues in device independence and then discuss the range of device independence architectures and technologies, which have been developed to address these. We finish with a description of our own DIWAF device independence framework.

8.4.1. Issues in Device Independence

The generation of content, and its subsequent delivery and presentation to a user is an involved process, and the problem of device independence can be viewed in a number of dimensions.

8.4.1.1. Separation of Concerns

Historically, the generation of the content of a document and the generation of its representation would have been handled as entirely separate functions. Authors would deliver a manuscript to a publisher, who would typeset the manuscript for publication. The skill of the typesetter was to make the underlying structure of the text clear to readers by consistent use of fonts, spacing and margins.

With the widespread availability of computers and word processors, authors often became responsible for both content and presentation. This blurring creates problems in device independent content delivery where content needs to be adapted to the device at hand, whereas much content produced today has formatting information embedded within it.

8.4.1.2. Location of Content Adaptation

Because of the client/server nature of web applications there are at least three distinct places where the adaptation of content to the device can occur:

Client Side Adaptation: all computer applications that display information to the user must have a screen driver that takes some internal representation of the data and transforms it into an image on the screen. In this sense, the client software is ultimately responsible for the presentation to the user. In an ideal world, providers would agree on a common data representation language for all devices, delegating responsibility for its representation to the client device. However, there are several mark-up languages in common use, each with a number of versions and variations, as well as a number of client side scripting languages. Thus the goal of producing a single universal representation language has proved elusive.

Server Side Adaptation: whilst the client is ultimately responsible for the presentation of data to the user, the display is driven by the data received from the server. In principle, if the server can identify the capabilities of the device being used, different representations of the content can be sent, according to the requirements of the client.

Because of the plethora of different data representations and device capabilities this approach has received much attention. A common approach is to define a data representation specifically designed to support device independence. These representations typically encourage a highly structured approach to content, achieve separation of content from style and layout, allow selection of alternative content and define an abstract representation of user interactions. In principle, these representations could be rendered directly on the client, but a pragmatic approach is to use this abstract representation to generate different presentations on the server.

Network Transformation: one of the reasons for the development of alternative data representations is the different network constraints placed upon mobile and fixed end-user devices. Thus a third possibility for content adaptation is to introduce an intermediate processing step between the server and client, within the network itself. For example, the widely used WAP protocol relies on a WAP gateway to transform bulky textual representations into compact binary representations of data. Another frequent application is to transform high-resolution colour images into low-resolution black and white.

8.4.1.3. Delivery Context

So far the discussion has focussed on the problems associated with using different hardware and software to generate an effective display of a single resource. However, this can be seen as part of a wider attempt to make web applications *context aware*.

Accessibility has been a concern to the W3C for a number of years, and in many ways the issues involved in achieving accessibility are parallel to the aims of achieving device independence. It may be, for example, that a user has a preference for using voice rather than a keyboard and from the point of view of the software, it is irrelevant whether this is because the device is limited, or because the user finds it easier to talk than type, or whether the user happens to need their hands for something else (e.g., to drive). To a large extent, any solutions developed for device independence will increase accessibility and vice versa.

Location is another important facet of context: a user looking for the nearest hotel will want to receive a different response depending on their current position.

User Profiles aim to enable a user to express and represent preferences about the way they wish to receive content—for example as text only, or in large font, or as voice XML. The Composite Capability/Preference Profile (CC/PP) standard (discussed in the next subsection) has been designed explicitly to take user preferences into consideration.

8.4.1.4. Device Identification

If device independence is achieved by client side interpretation of a universal mark-up language, then identification of device capabilities can be built into the browser. However, if the server transformation model is taken, then there arises the problem of identifying the target device from the HTTP request.

Two approaches to this problem have emerged as common solutions. The current W3C recommendation is to use CC/PP (Klyne, 2004), a generalisation of the UAProf standard developed by the Wireless Application Protocol Forum (now part of the Open Mobile Alliance) (WAPF, 1999). In this standard, devices are described as a collection of components, each with a number of attributes. The idea is that manufacturers will provide profiles of their devices, which will be held in a central device repository. The device will identify itself using HTTP Header extensions, enabling the server to load its profile. One of the strengths of this approach is that users (or devices, or network elements) are able to specify to the default device data held centrally on a request-by-request basis. Another attraction of the specification is that it is written in RDF (MacBride, 2004), which makes it easy to assimilate into a larger ontology, for example including user profiles. The standard also includes a protocol, designed to access the profiles over low bandwidth networks.

An alternative approach is the Wireless Universal Resource File (WURFL) (Passani, 2005). This is a single XML document, maintained by the user community and freely available, containing a description of every device known to the WURFL community (currently around 5000 devices). The aim is to provide an accurate and up to date characterisation of wireless devices. It was developed to overcome the difficulty that manufacturers do not always supply accurate CC/PP descriptions of their devices. Devices are identified using the standard user-agent string sent with the request. The strength of this approach is that devices are arranged in an inheritance hierarchy, which means that sensible defaults can be inferred even if only the general class of device is known. CC/PP and WURFL are described in more detail later in this section.

8.4.2. Device Independence Architectures and Technologies

The rapid advance of mobile communications has spurred numerous initiatives to bridge the gap between existing fixed PC technologies and the requirements of mobile devices. In particular, the World Wide Web Consortium (W3C) has a number of active working groups, including the Device Independence Working Group, which has produced a range of material on this issue.[6] In this section, we give an overview of some of the more prominent device independence technologies.

8.4.2.1. XFORMS

XForms (Raman, 2003) is an XML standard for describing web-based forms, intended to overcome some of the limitations of HTML. Its key feature is the separation of traditional forms into three parts—the data model, data instances and presentation. This allows a more natural expression of data flow and validation, and avoids many of the problems associated with the use of client side scripting languages. Another advantage is strong integration with other XML technologies such as the use of XPath to link documents.

XFORMS is not intended as a complete solution for device independence, and it does not address issues such as device recognition and content selection. However, its separation of the abstract data model from presentation addresses many of the issues in the area of user interaction, and the XFORMS specification is likely to have an impact on future developments.

8.4.2.2. CSS3 and Media Queries

Cascading Style Sheets is a technology which allows the separation of content from format. One of the most significant benefits of this approach is that it allows the 'look and feel' of an entire web site to be specified in a single document. CSS version 2 also provided a crude means of selecting content and style based on the target device using a 'media' tag.

CSS3 greatly extends this capability by integrating CC/PP technology into the style sheets, via Media Queries (Lie, 2002), allowing the user to write Boolean expressions which can be used to select different styles depending on attributes of the current device. In particular, content can be omitted altogether if required. Unfortunately, media queries do not yet enjoy consistent browser support.

---

[6]http://www.w3.org/2001/di/

8.4.2.3. XHTML-Mobile Profile

This is a client side approach to device independence. Its aim is to define a version of HTML which is suitable for both mobile and fixed devices. Issues to do with device capability identification and content transformation are bypassed, since the presentation is controlled by the browser on the client device. The XHTML mobile profile specification (WAPF, 2001) draws on the experience of WML and the compact HTML (cHTML) promoted by I-mode in Japan, and increasingly penetrating into Europe.

8.4.2.4. SMIL

The Synchronised Multi-media Integration Language (SMIL) (Butterman *et al.*, 2004) is another mark-up language for describing content. This time the focus is on multimedia, and in particular on animation, but the SMIL specification is very ambitious, and includes sophisticated models for describing layout and content selection. SMIL is perhaps currently the most complete specification language for server-side transformation. However, there does not yet seem to have been significant take up in the device independence arena.

8.4.2.5. COCOON/DELI

Section 8.4.1.4 discussed the CC/PP protocol, which is the current W3C recommendation for device characterisation. A Java API has been developed for this protocol as an open source project by SUN, building on work done at HP under the name DELI (Jacobs and Jaj, 2005). This provides a simple programming interface to CC/PP which allows developers to access the capabilities of the current device. This has been integrated into COCOON,[7] a framework for building web resources using XML as the content source, and using XSLT to transform this into suitable content based on the current device.

A disadvantage of this approach is the effort required to write suitable XSLT style sheets.

8.4.2.6. WURFL/WALL

The Wireless Universal Resource File has been briefly described in Section 8.4.1.4. One of the most useful features of the WURFL is its hierarchical structure; devices placed at lower nodes in the tree inherit the properties of their ancestors. This gives the WURFL a certain degree

---

[7]http://cocoon.apache.org/

of robustness against additions. Even if a device cannot be located in the file, default values can be assumed from its 'family', inferred from its manufacturer and series number.

The WURFL claims to have greater take up than the CC/PP standard, and its reliability, accuracy and robustness are attractive features. However, it has certain disadvantages. In particular, it does not provide any information about the network, the software or user preferences. An ideal solution would be recast the WURFL in RDF so that it could be integrated with CC/PP. However, RDF does not support inheritance, the WURFL's key advantage.

In order to make the WURFL accessible to developers, OpenWave have developed APIs in Java and PHP that provide a simple programming interface. They have also developed a set of java tag libraries, for use in conjunction with Java Server Pages (JSP), known as WALL.[8] WALL appears to be the closest approach yet to the ideal of device independence. Using WALL it is possible to write a single source, in a reasonably intuitive language, which will result in appropriate content being delivered to the target device without any further software development.

8.4.3. DIWAF

The SEKT Device Independence Web Application Framework is a server side application which provides a framework for presenting structured data to the user (Glover and Davies, 2005). The framework does not use a mark-up language to annotate the data. Instead, it makes use of templates, which are 'filled' with data rather like a mail merge in a word processor. These templates allow the selection, repetition and rearrangement of data, interspersed with static text. The framework can select different templates according to the target device, which present the same data in different ways.

The approach is some ways analogous to XSLT. Data is held internally structured according to some logical business model. This data can be selected and transformed into a suitable presentation model by the framework. However, there are some significant advantages of this approach over XSLT. First the data source does not have to be an XML document, but may be a database or structured text file. Second the templates themselves do not have to be XML documents. This means that they can be designed using appropriate tools—for example HTML documents can be written using an HTML editor. Finally, the templates are purely declarative and contain no programming constructs. This means that no special technical knowledge is required to produce them.

---

[8]http:// developer.openwave.com

Very often effective presentations can be produced directly from the logical data model. However, sometimes the requirements go beyond the capabilities of declarative templates. For example it may be necessary to perform calculations or text processing. For this reason, the framework has a three tier, Model-View-Control architecture. The first layer is the logical data model. The second layer contains the business logic which performs any necessary processing. The third and final layer is the presentation layer where the data is transformed into a suitable format for presentation on the target device. This architecture addresses the separation of concerns issue discussed in Section 8.4.1.1.

In the current implementation of the DIWAF, device identification uses the RDF-based CC/PP (an open standard from W3C), with an open source Java implementation. In this framework, device profile information is made available to Java servlets as a collection of attributes, such as screen size, browser name etc. These attributes can be used to inform the subsequent selection and adaptation of content, by combining them in Boolean expressions. Figure 8.4 shows exactly the same content (located

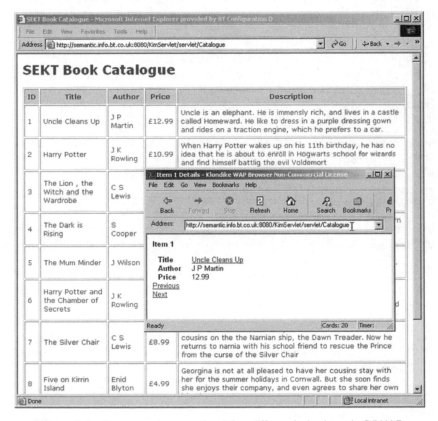

Figure 8.4 Repurposing content for different devices in DIWAF.

at the same URL) rendered via DIWAF on a standard web browser and on a WAP browser emulator.

We have used this framework to support delivery of knowledge to users on a variety of devices in the SEKTAgent system, as discussed later in this chapter. Further details of this approach are available in Glover and Davies (2005).

8.5. SEKTAGENT

We have seen in Section 8.2 how some semantic search tools use an ontological knowledge base to enhance their search capability. We discussed in Sections 8.3 and 8.4 the use of natural language generation to describe ontological knowledge in a more natural format and the delivery of knowledge to the user in a format appropriate to the terminal device to which they currently have access. In this section, we describe a semantic search agent, SEKTAgent, which brings together the exploitation of an ontological knowledge base, natural language generation and device independence to proactively deliver relevant information to its users.

Search agents can reduce the overhead of completing a manual search for information. The best known commercial search agent is perhaps 'Google Alerts', based on syntactic queries of the Google index.

Using an API provided by the KIM system (see Section 8.2.5 above), SEKTAgent allows users to associate with each agent a *semantic* query based upon the PROTON ontology (see Chapter 7). Some examples of agent queries that could be made would be for documents mentioning:

- A named person holding a particular position, within a certain organisation.
- A named organisation located at a particular location.
- A particular person and a named location.
- A named company, active in a particular industry sector.

This mode of searching for types of entity can be complemented with a full text search, allowing the user to specify terms which should occur in the text of the retrieved documents.

In addition to the use of subsumption reasoning provided by KIM, it is also planned that SEKTAgent will incorporate the use of explicitly defined domain-specific rules. The SEKT search tool uses KAON2[9] as its reasoning engine. KAON2 is an infrastructure for managing OWL-DL ontologies. It provides an API for the programmatic management of OWL-DL and an inference engine for answering conjunctive queries expressed using SPARQL[10] syntax.

---

[9]http://kaon2.semanticweb.org/
[10]http://www.w3.org/TR/rdf-sparql-query/

KAON2 allows new knowledge to be inferred from existing, explicit knowledge with the application of rules over the ontology. Consider a semantic query to determine who has collaborated with a particular author on a certain topic. This query could be answered through the existence of a rule of the form:

```
If (?personX isAuthorOf ?document) & (?personY isAuthorOf
?document) -> (?personX collaboratesWith ?personY) &
(?personY collaboratesWith ?personX)
```

This rule states that if two people are authors of the same document then they are collaborators. When a query involving the collaborateswith predicate is submitted to KAON2, the above rule is enforced and the resulting inferred knowledge returned as part of the query.

Figure 8.5 illustrates the results page for an agent which is searching for a person named 'Ben Verwaayen' within the organisation 'BT'. SEKTAgent is automatically run offline[11] at a periodicity specified by the user (daily, weekly etc.). When new results (i.e. ones not previously

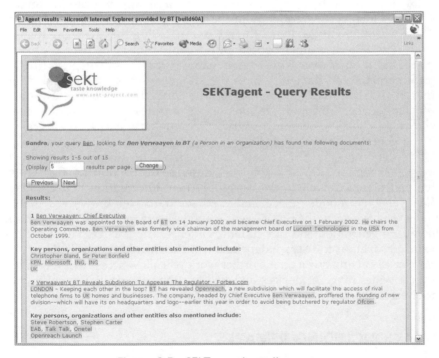

Figure 8.5 SEKTagent results page.

---

[11]Offline in this context means automatically without any user interaction.

Microsoft Corporation is a Public Company located in United States and Worldwide. Designs, develops, manufactures, licenses, sells and supports a wide range of software products. Its webpage is www.microsoft.com. It is traded on NASDAQ with the index MSFT. Key people include:

- Bill Gates – Chairman, Founder

- Steve Balmer – CEO

- John Conners – Chief Finanacial Officer

Last year its revenues were $36.8bn and its net income was $8.2bn.

Figure 8.6   ONTOSUM generated description.

presented by the agent to the given user) satisfying this query are found, the user is sent a message which includes a link to an agent results page. For each result found, the title of the page and a short summary of the content relevant to the query are displayed. The summary highlights the occurrences of the named entities that satisfy the query. Other recognised named entities are also highlighted and the class to which each entity belongs is shown by a colour coding scheme. Following the summary, entities which occur frequently in the result documents are also shown. These are other entities that although not matching the query are related to it — in this case other people and organisations. The user is able to place his mouse over any of the named entities to display further information about the entity from the knowledge base, generated using the ONTOSUM NLG system described in Section 8.3. For example, mousing over 'Microsoft' in the list of entities in the results page shown in Figure 8.5 would result in the summary shown in Figure 8.6 being generated by ONTOSUM.

Results from the SEKTAgent can be made available via multiple devices using the DIWAF framework described in Section 8.4. Currently, templates are available to deliver SEKTAgent information to users via a WAP-enabled mobile device, and via a standard web browser.

As we have seen, the SEKTAgent combines semantic searching, natural language generation and device independence to proactively deliver relevant information to users independent of the device to which they may have access at any given time. Further work will allow access to information over a wider range of devices and will test the use of SEKTAgent in real user scenarios, such as that described in Chapter 11 of this volume.

8.6. CONCLUDING REMARKS

The current means of knowledge access for most users today is the traditional search engine, whether searching the public Web or the corporate intranet. In this chapter, we began by identifying and discussing some shortcomings with current search engine technology. We then described how the use of semantic web technology can address some of these issues. We surveyed research in three areas of knowledge access: the use of ontologies and associated metadata to enhance searching and browsing; the generation of natural language text from such formal structures to complement and enhance semantic search; and the delivery of knowledge to users independent of the device to which they have access. Finally, we described SEKTAgent, a research prototype bringing together these three technologies into a semantic search agent. SEKTAgent provides an early glimpse of the kind of semantic knowledge access tools which will become increasingly commonplace as deployment of semantic web technology gathers pace.

REFERENCES

Aguado G, Bañón A, Bateman JA, Bernardos S, Fernández M, Gómez-Pérez A, Nieto E, Olalla A, Plaza R, Sánchez A. 1998. *ONTOGENERATION: Reusing domain and linguistic ontologies for Spanish text generation.* Workshop on Applications of Ontologies and Problem Solving Methods, ECAI'98.

Bateman JA, Magnini B, Fabris G. 1995. *The Generalized Upper Model Knowledge Base: Organization and Use.* Towards Very Large Knowledge Bases, pp 60–72.

Bernstein A, Kaufmann E, Goehring A, Kiefer C. 2005. *Querying Ontologies: A Controlled English Interface for End-users.* In *Proceedings 4th International Semantic Web Conference, ISWC2005,* Galway, Ireland, November 2005, Springer-Verlag.

Bontcheva K, Wilks Y. 2004. *Automatic Report Generation from Ontologies: The MIAKT approach.* Ninth International Conference on Applications of Natural Language to Information Systems (NLDB'2004).

Brin S, Page L. 1998. *The anatomy of a large-scale hypertextual web search engine.* In *proceedings of the 7th International World Wide Web Conference,* Brisbane, Australia, pp 107–117.

de Bruijn J, Martin-Recuerda F, Manov D, Ehrig M. 2004. *State-of-the-art survey on Ontology Merging and Aligning v1.* Technical report, SEKT project deliverable 4.2.1. http://sw.deri.org/jos/sekt-d4.2.1-mediation-survey-final.pdf.

Buchanan BG, Moore JD, Forsythe DE, Cerenini G, Ohlsson S, Banks G. 1995. *An intelligent interactive system for delivering individualized information to patients. Artificial Intelligence in Medicine* 7:117–154.

Butterman D, Rutledge L. 2004. *SMIL 2.0: Interactive Multimedia for Web and Mobile devices.* Springer-Verlag Berlin and Heidelberg GmbH & Co. K.

Cohen S, Kanza Y, Kogan Y, Nutt W, Sagiv Y, Serebrenik A. 2002. EquiX: A search and query language for *XML. Journal of the American Society for Information Science and Technology,* 53(6):454–466.

Cohen S, Mamou J, Kanza Y, Sagiv S. 2003. *XSEarch: A Semantic Search Engine for XML.* In *proceedings of the 29th VLDB Conference,* Berlin, Germany.

Dale R, Oberlander J, Milosavljevic M, Knott A. 1998. Integrating natural language generation and hypertext to produce dynamic documents. *Interacting with Computers* 11:109–135.

Davies J. 2000. *QuizXML: An XML search engine,* Informer, Vol 10, Winter 2000, ISSN 0950-4974, http://irsg.bcs.org/informer/Winter_00.pdf.

Davies J, Bussler C, Fensel D, Studer R (eds). 2004. *The Semantic Web: Research and Applications*. In *Proceedings of ESWS 2004*, LNCS 3053, Springer-Verlag, Berlin.

Davies J, Fensel D, van Harmelen F. 2003. *Towards the Semantic Web*. Wiley; UK.

Davies J, Weeks R, Krohn U, *QuizRDF: Search technology for the semantic web*, in (Davies *et al.*, 2003).

Ding L, Finin T, Joshi A, Pan R, Cost RS, Peng Y, Reddivari P, Doshi V, Sachs J. 2004. *Swoogle: A Search and Metadata Engine for the Semantic Web*, Conference on Information and Knowledge Management CIKM04, Washington DC, USA, November 2004.

Domingue J, Dzbor M, Motta E. 2004. *Collaborative Semantic Web Browsing with Magpie* in (Davies *et al.*, 2004).

Fensel D, Studer R (eds): 2004. In *Proceedings of ESWC 2004*, LNCS 3053, Springer-Verlag, pp 388–401.

Florescu D, Kossmann D, Manolescu I. 2000. Integrating keyword search into XML query processing. *The International Journal of Computer and Telecommunications Networking* 33(1):119–135.

Glaser H, Alani H, Carr L, Chapman S, Ciravegna F, Dingli A, Gibbins N, Harris S, Schraefel MC, Shadbolt N. *CS AKTiveSpace: Building a Semantic Web Application* in (Davies *et al.*, 2004).

Glover T, Davies J. 2005. Integrating device independence and user profiles on the web. *BT Technology Journal* 23(3):JXX.

Grčar M, Mladenić D, Grobelnik M. 2005. *User profiling for interest-focused browsing history*, SIKDD 2005, http://kt.ijs.si/dunja/sikdd2005/, Slovenia, October 2005.

Guha R, McCool R. 2003. Tap: A semantic web platform. *Computer Networks* 42:557–577.

Guha R, McCool R, Miller E. 2003. *Semantic Search*. WWW2003, 20–24 May, Budapest, Hungary.

Guo L, Shao F, Botev C, Shanmugasundaram J. 2003. *XRANK: Ranked Search over XML Documents*. SIGMOD 2003, June 9–12, San Diego, CA.

Hoh S, Gilles S, Gardner MR., 2003. Device personalisation—Where content meets device. *BT Technology Journal* 21(1):JXX.

Huynh D, Karger D, Quan D. 2002. *Haystack: A Platform for Creating, Organizing and Visualizing Information Using RDF. In proceedings of the WWW2002 International Workshop on the Semantic Web, Hawaii, 7 May 2002*.

Iosif V, Mika P, Larsson R, and Akkermans H. 2003. *Field Experimenting with Semantic Web Tools in a Virtual Organisation* in Davies (2003).

Jacobs N, Jaj J. 2005. *CC/PP Processing*. Java Community Process JSR-000188, http://jcp.org/aboutJava/communityprocess/final/jsr188/index.html, accessed 21/11/2005.

Jansen BJ, Spink A, Saracevic T. 2000. Real life, real users, and real needs: A study and analysis of user queries on the web. *Information Processing and Management* 36(2):207–227.

Klyne G *et al.* (editors). 2004. *CC/PP Structure and Vocabularies*. W3C Recommendation 15 Jan 2004. http://www.w3.org/TR/

Li J, Pease A, Barbee C, *Experimenting with ASCS Semantic Search*, http://reliant.teknowledge.com/DAML/DAML.ps. Accessed on 9 November 2005.

Lie HW. *et al.* 2002. *Media Queries*. W3C Candidate Recommendation 2002, available at http://www.w3.org/TR/2002/CR-css3-mediaqueries-20020708/.

MacBride B. (Series Editor). 2004. *Resource Description Framework (RDF) Syntax Specification*. W3C Recommendation (available at www.w3.org/TR/rdf-syntax-grammar/).

Passani L, Trasatti A. 2005. *The Wireless Universal Resource File*. Web Resource http://wurfl.sourceforge.net, accessed 21/11/2005.

Popov B, Kiryakov A, Kirilov A, Manov D, Ognyanoff D, Goranov M. 2003. *KIM— Semantic Annotation Platform* in 2nd International Semantic Web Conference (ISWC2003), 20–23 October 2003, Florida, USA. LNAI Vol. 2870, Springer-Verlag Berlin Heidelberg pp 834–849.

Quan D, Karger DR. 2004. *How to Make a Semantic Web Browser*. The Thirteenth International World Wide Web Conference, New York City, 17–22 May IW3C2, ACM Press.

Raman TV. 2003. *XForms: XML Powered Web Forms*. Addison Wesley: XX.

Reiter E, Dale R. 2000. *Building Natural Language Generation Systems*. Cambridge University Press, Cambridge.

Resource Description Framework (RDF): Concepts and Abstract Syntax, W3C Recommendation 10 February 2004, http://www.w3.org/TR/2004/REC-rdf-concepts-20040210

Rocha C, Schwabe D, de Aragao MP. 2004. *A hybrid approach for searching in the semantic web*. WWW 2004, 17–22 May, New York, USA.

Salton G, Wong A, Yang CS. *A Vector Space Model for Automatic Indexing*, in [Sparck-Jones and Willett, 1997]

Sparck-Jones, K, Willett P. 1997. *Readings in Information Retrieval*, Morgan-Kaufman: California, USA.

Spink A, Jansen BJ, Wolfram D, Saracevic T. 2002. From E-Sex to E-Commerce: Web Search Changes. *Computer XX*: 107–109.

Vallet D, Fernandez M, Castells P. 2005. An *Ontology-based Information Retrieval Model*. In Proceedings 2nd European Semantic Web Conference, ESWC2005, Heraklion, Crete, May/June 2005, Springer-Verlag, Berlin.

WAPF (Wireless Application Protocol Forum). 1999. *User Agent Profile Specification*. http://www.openmobilealliance.org

WAPF (Wireless Application Protocol Forum). 2001. XHTML *Mobile Profile*. http://www.openmobilealliance.org

Wilcock G. 2003. *Talking OWLs: Towards an Ontology Verbalizer*. Human Language Technology for the Semantic Web and Web Services, ISWC'03, Sanibel Island, Florida, pp 109–112.

Wilcock G, Jokinen K. 2003. *Generating Responses and Explanations from RDF/XML and DAML + OIL*. Knowledge and Reasoning in Practical Dialogue Systems, IJCAI-2003 Acapulco, pp 58–63.

9

Ontology Engineering Methodologies

York Sure, Christoph Tempich and Denny Vrandecic

9.1. INTRODUCTION

The two main drivers of practical knowledge management are technology and people, as pointed out earlier by Davenport (1996). Traditional IT-supported knowledge management applications are built around some kind of corporate or organizational memory (Abecker *et al.*, 1998). Organizational memories integrate informal, semi-formal, and formal knowledge in order to facilitate its access, sharing, and reuse by members of the organization(s) for solving their individual or collective tasks (Dieng *et al.*, 1999), for example as part of business processes (Staab and Schnurr, 2002).

The knowledge structures underlying such knowledge management systems constitute a kind of ontology (Staab and Studer, 2004) that may be built according to established methodologies (Fernandez-Lopez *et al.*, 1999; Sure, 2003). These methodologies have a centralized approach towards engineering knowledge structures requiring *knowledge engineers*, *domain experts*, and others to perform various tasks such as a *requirement analysis* and *interviews*. While the user group of such an ontology may be huge, the development itself is performed by a—comparatively—small group of domain experts who *provide the model for the knowledge*, and ontology engineers who *structure and formalize* it.

Decentralized knowledge management systems are becoming increasingly important. The evolving Semantic Web (Berners-Lee *et al.*, 2001)

Semantic Web Technologies: Trends and Research in Ontology-based Systems
John Davies, Rudi Studer, Paul Warren © 2006 John Wiley & Sons, Ltd

will foster the development of numerous use cases for this new paradigm. Therefore, methodologies based on traditional, centralized knowledge management systems are no longer feasible. There are some technical solutions toward Peer-to-Peer knowledge management systems (e.g., Bonifacio *et al.*, 2003; Ehrig *et al.*, 2003). Still, the traditional methodologies for creating and maintaining knowledge structures appear to be unusable in distributed and decentralized settings, and so the systems that depend on them will fail to cope with the dynamic requirements of big or open user groups.

The chapter is structured as follows. First, we define *methodology* and *ontology engineering methodology* in Section 9.2. Then we provide a survey of existing ontology engineering methodologies in Section 9.3. Since we believe that the engineering of ontologies in practical settings requires tool support to cope with the various complex tasks we also include a survey of corresponding ontology engineering tools in this section. The survey ends with an enumeration of open research issues. We partly address these research issues with the new DILIGENT (*Distributed, Loosely controlled, and evolvInG Engineering of oNTologies*) methodology which is introduced in Section 9.4. Before concluding we present some first lessons learned from applying DILIGENT in a number of case studies.

9.2. THE METHODOLOGY FOCUS

It has been a widespread conviction in knowledge engineering that methodologies for building knowledge-based systems help knowledge engineering projects to successfully reach their goals in time (cf. Schreiber *et al.*, 1999 for one of the most widely deployed methodologies). With the arrival of ontologies in knowledge-based systems the same kind of methodological achievement for structuring the ontology-engineering process has been pursued by approaches like the ones presented in the next section.

In this section, we will look at the general criteria for and specific properties of methodologies for the ontology life cycle. We will first apply a definition of *methodology* to our field of interest, and then point out to the conclusions drawn from this definition.

9.2.1. Definition of Methodology for Ontologies

The IEEE defines a methodology as 'a comprehensive, integrated series of techniques or methods creating a general systems theory of how a class of thought-intensive work ought be performed' (IEEE, 1990). A methodology should define an 'objective (ideally quantified) set of criteria for determining whether the results of the procedure are of acceptable quality.'[1]

By contrast, a *method* is a 'orderly process or procedure used in the engineering of a product or performing a service' (IEEE, 1990). A *technique* is 'a technical and managerial procedure used to achieve a given objective' (IEEE, 1990).

A *process* is a 'function that must be performed in the software life cycle. A process is composed of activities' (IEEE, 1996). An *activity* is 'a constituent task of a process' (IEEE, 1996). A *task* 'is a well defined work assignment for one or more project members. Related tasks are usually grouped to form activities' (IEEE, 1996).

9.2.2. Methodology

An ontology engineering methodology needs to consider the following three types of activities:

- Ontology management activities.
- Ontology development activities.
- Ontology support activities.

9.2.2.1. Ontology Management Activities

Procedures for ontology management activities must include definitions for the scheduling of the ontology engineering task. Further it is necessary to define control mechanism and quality assurance steps.

9.2.2.2. Ontology Development Activities

When developing the ontology it is important that procedures are defined for environment and feasibility studies. After the decision to build an ontology the ontology engineer needs procedures to specify, conceptualize, formalize, and implement the ontology. Finally, the users and engineers need guidance for the maintenance, population, use, and evolution of the ontology.

9.2.2.3. Ontology Support Activities

To aid the development of an ontology, a number of important supporting activities should be undertaken. These include knowledge acquisition, evaluation, integration, merging and alignment, and configuration management. These activities are performed in all steps of the development and management process. Knowledge acquisition can happen in a centralized as well as a decentralized way. Ontology

---

[1] http://computing-dictionary.thefreedictionary.com/Methodology

learning is a way to support the manual knowledge acquisition with machine learning techniques.

9.2.3. Documentation

It is important to document the results after each activity. In a later stage of the development process this helps to trace why certain modeling decisions have been undertaken. The documentation of the results can be facilitated with appropriate tool support. Depending on the methodology the documentation level can be quite different. One methodology might require documenting only the results of the ontology engineering process while others give the decision process itself quite some importance.

9.2.4. Evaluation

In the ontology engineering setting, evaluation measures should provide means to measure the quality of the created ontology. This is particular difficult for ontologies, since modeling decisions are in most cases subjective. A general survey of evaluation measures for ontologies can be found in Gomez-Perez (2004). Additionally we want to refer to the evaluation measures which can be derived from statistical data (Tempich and Volz, 2003) and measures which are derived from philosophical principles. One of the existing approaches for ontology evaluation is OntoClean (Guarino and Welty, 2002).

9.3. PAST AND CURRENT RESEARCH

In the following we summarize the distinctive features of the available ontology engineering methodologies and give quick pointers to existing tool support specifically targeted to the methodologies. Next, we briefly introduce the most prominent existing tools.

9.3.1. Methodologies

An extensive state-of-the-art overview of methodologies for onto-logy engineering can be found in Gomez-Perez et al. (2003). More recently Cristani and Cuel (2005) proposed a framework to compare ontology engineering methodologies and evaluated the established ones accordingly. In the OntoWeb[2] project, the members gathered guide-lines and best practices for industry (Leger et al., 2002a, b) with a focus on

---

[2] see http://www.ontoweb.org/

applications for E-Commerce, Information Retrieval, Portals and Web Communities. A very practical oriented description to start building ontologies can be found in Noy and McGuinness (2001).

In our context, the following approaches are especially noteworthy. Where it is adequate we give pointers to tools mentioned in the next section, whenever tool support is available for a methodology.

CommonKADS (Schreiber *et al.*, 1999) is not *per se* a methodology for ontology development. It covers aspects from corporate knowledge management, through knowledge analysis and engineering, to the design and implementation of knowledge-intensive information systems. CommonKADS has a focus on the initial phases for developing knowledge management applications. CommonKADS is therefore used in the OTK methodology for the early feasibility stage. For example, a number of worksheets can be used to guide a way through the process of finding potential users and scenarios for successful implementation of knowledge management. CommonKADS is supported by PC PACK, a knowledge elicitation tool set, that provides support for the use of elicitation techniques such as interviewing, that is it supports the collaboration of knowledge engineers and domain experts.

DOGMA (Jarrar and Meersman, 2002; Spyns *et al.*, 2002) is a database-inspired approach and relies on the explicit decomposition of ontological resources into *ontology bases* in the form of simple binary facts called lexons and into so-called ontological commitments in the form of description rules and constraints. The modeling approach is implemented in the DOGMA Server and accompanying tools such as the DOGMA Modeler tool set.

The *Enterprise Ontology* (Uschod and King, 1995; Uschold *et al.*, 1998) proposed three main steps to engineer ontologies: (i) to identify the purpose, (ii) to capture the concepts and relationships between these concepts, and the terms used to refer to these concepts and relationships, and (iii) to codify the ontology. In fact, the principles behind this methodology influenced much work in the ontology community. Explicit tool support is given by the Ontolingua Server, but actually these principles heavily influenced the design of most of today's more advanced ontology editors.

The *KACTUS* (Bernaras *et al.*, 1996) approach requires an existing knowledge base for the ontology development. The ontology is build based on the existing knowledge model, applying a bottom-up strategy. There is no specific tool support for this methodology.

METHONTOLOGY (Fernandez-Lopez *et al.*, 1999) is a methodology for building ontologies either from scratch, reusing other ontologies as they are, or by a process of re-engineering them. The framework enables the construction of ontologies at the 'knowledge level,' that is the conceptual level, as opposed to the implementation level. The framework consists of: identification of the ontology development process containing the main activities (evaluation, configuration, management, conceptualization, integration implementation, etc.); a lifecycle based on evolving proto-types; and the methodology itself, which specifies the steps to be taken to perform each activity, the techniques used, the products to be output and how they are to be evaluated. METHONTOLOGY is partially supported by WebODE (Arpirez *et al.*, 2001).

SENSUS (Swartout *et al.*, 1997) is a top-down and middle-out approach for deriving domain specific ontologies from huge ontologies. The methodology is very specialized and does not cover the engineering of ontologies as such.

TOVE (Uschold and Grueninger, 1996) proposes a formalized method for building ontologies based on competency questions. The approach of using competency questions, which describe the questions that an ontology should be able to answer, has been proven to be very helpful in practical scenarios especially when dealing with domain experts with little knowledge in modeling.

HOLSAPPLE (Holsapple and Joshi, 2002) proposes a methodology for collaborative ontology engineering. The aim of their work is to support the creation of a static ontology. A knowledge engineer defines an initial ontology which is extended and changed based on the feedback from a panel of domain experts. The feedback is collected with a questionnaire. The knowledge engineer examines the questionnaires, incorporates the new requirements and a new questionnaire is send around, until all participants agree with the outcome.

HCONE (Kotis and Vouros, 2003; Kotis *et al.*, 2004) is a recent approach to ontology development. HCONE stands for Human Centered ONto-logy Environment. It supports the development of ontologies in a decentralized fashion. Three different spaces are introduced in which ontologies can be stored. The first one is the *Personal Space*. In this space users can create and merge ontologies, control ontology versions, map terms and word senses to concepts, and consult the top level ontology. The evolving personal ontologies can be shared in the *Shared Space*. The shared space can be accessed by all participants. In the shared space users can discuss ontological decision. After a discussion and agreement the ontology is moved to the *Agreed space*.

The *OTK Methodology* (Sure, 2003) was developed in the EU project On-To-Knowledge. The OTK Methodology divides the ontology engineering task into five main steps. Each step has numerous sub-steps, requires a main decision to be taken at the end and results in a special outcome. The steps are 'Feasibility Study,' 'Kickoff,' 'Refinement,' 'Evaluation,' and 'Application and Evolution.' Within the steps there are sub-steps for example 'Refinement' has sub-steps which include 'Refine semi-formal ontology description,' 'Formalize into target ontology,' and 'Create prototype.' Documents resulting from each phase, for example for the 'Kickoff' phase an 'Ontology Requirements Specification Document (ORSD)' and the 'Semi-formal ontology description' are created. The documents are the basis for the major decisions that have to be taken at the end to proceed to the next phase, for example whether in the 'Kickoff' phase one has captured sufficient requirements. The major outcomes typically serve as support for the decisions to be taken. The phases 'Refinement—Evaluation—Application—Evolution' typically need to be performed in iterative cycles. One might notice that the development of such an application is also driven by other processes, for example software engineering and human issues. All steps of the methodology are supported by plugins available for the tool OntoEdit (Sure *et al.*, 2002, 2003). In a nutshell, the OTK Methodology completely describes all steps which are necessary to build an ontology for a centralized system.

UPON, the Unified Process for ONtology building, has been proposed in Nicola *et al.* (2005). Although the methodology has not been well tested in projects yet, and tool support is still in its infancy, it is conceptually well founded. It is based on the Unified Software Development Process and supported by UML (Unified Modeling Language). UPON defines a series of work flows which are cyclically performed in different phases. The work flows are: (1) requirements identification, for example by writing a story board and using competency questions, (2) analysis, which includes the identification of existing resources and the modeling of the application scenario, (3) design and conceptualization, (4) implementation, and (5) test in which the coverage of the application domain should be guaranteed and the competency questions are evaluated. The work flows are followed in the four phases: (1) inception, (2) elaboration, (3) construction, and (4) transition defined in the methodology. These phases are performed in a cyclic manner. After each cycle an applicable ontology is produced.

For the sake of completeness and without a detailed description we here reference some other proposals for structured ontology engineering. Among them are Pinto and Martings (2001), advocating an approach of ontology building by reuse. One of their major findings was that current methodologies offer only limited support for axiom building although it is a part of ontology engineering which takes a lot of time. In Gangemi *et al.* (1998), the authors outline the ONIONS approach. ONIONS (ONtologic Integration Of Naive Sources) creates a common framework to generalize and integrate the definitions that are used to organize a set of terminological sources. In other words, it allows the coherent development of a terminological domain ontology (a terminological ontology is usually defined as the explicit conceptualization of a vocabulary) for each source, which can be then compared with the others and mapped to an integrated ontology library.

9.3.2. Ontology Engineering Tools

An early overview of tools that support ontology engineering can be found in Duineveld *et al.* (2000). Joint efforts of members of the thematic

network OntoWeb have also provided an extensive state-of-the-art overview on ontology related tools, including Ontology Engineering Environments (OEE) (Gomez-Perez *et al.*, 2002). An evaluation of ontology
engineering environments has been performed as part of the EON 2002
workshop (Sure and Angele, 2002). In our context, the following tools are
especially noteworthy.

KAON OImodeller (Bozsak *et al.*, 2002; Motik *et al.*, 2002) belongs to the
KAON tool suite. The system is designed to be highly scalable and relies
on an advanced conceptual modeling approach that balances some
typical trade-offs to enable a more easy integration into an existing
enterprise information infrastructure. Recently the OWL-DL and SWRL
(Semantic Web Rule Language[3]) reasoning engine KAON2[4] has been
added to the KAON landscape of tools. The extension of the KAON tool
suit is part of the SEKT project.

Protégé (Noy *et al.*, 2000) is a well-established ontology editor with a
large user community. It was the first editor with an extensible plug-in
structure and it relies on the frame paradigm for modeling. Numerous
plug-ins from external developers exist. It also supports current standards like RDF(S) and OWL. Recently support for axioms was added
through the 'PAL tab' (*Protégé* axiom language, cf. Hou *et al.*, 2002).

WebODE (Arpirez *et al.*, 2001) is an 'ontology engineering workbench'
that provides various services for ontology engineering. It is accompanied by a sophisticated methodology of ontology engineering, *viz.*
METHONTOLOGY. WebODE is purely web-based and is built on top
of an application server. For inferencing services it relies on Prolog. It
provides basic translators to current standards such as RDF(S) and OWL.

OntoEdit (Sure *et al.*, 2002, 2003) supports explicitly the OTK Methodology. The open plug-in framework enables the integration of a number
of extensions to the basic ontology management services OntoEdit
provides. In particular OntoEdit offers advanced support for collaboration and integration of the inferencing capabilities. A reimplementation
of OntoEdit based on the Eclipse[5] framework has recently been made
available under the new name *OntoStudio*.

9.3.3. Discussion and Open Issues

We have surveyed a number of ontology engineering methodologies and
their main strengths. We conclude that none of the existing methodologies cover all aspects of ontology engineering and that there still exist
many open issues. Most of the methodologies address the engineering of

---

[3] http://www.w3.org/Submission/SWRL/
[4] http://kaon2.semanticweb.org/
[5] http://www.eclipse.org/

a single ontology for a particular application and do not support maintenance activities after the first release. The methodologies proposed more recently do treat the evolution of the ontology seriously. However, while early phases of the development process are well understood and detailed activity descriptions exists (e.g., how to create competency questions), more fine-grained guidelines for later stages in the process are still missing. For example, methodological support for the creation and evolution of complex logical axioms is still an open issue. The number of best practices describing concrete development efforts is still very small. With this approach the quality of the development and maintenance effort depends mainly on the capabilities of the actors involved. In particular for multisite development efforts no clear guidelines exist as here multiple views should be considered. Multiple views on the same domain can lead to different conceptualizations, making agreement on a shared one therefore particularly difficult.

Another important aspect of ontology building, namely the construction of ontologies with the help of automated methods is not directly supported by any of the existing methodologies. Although the quality of ontology learning methods with regards to the usability of results has increased tremendously in the past years, the selection of appropriate input information or the integration of the produced ontologies with manual ones is still not well understood.

The number of existing methodologies covering different aspects of the ontology building process, suggest taking a 'method engineering' approach as has happened in software engineering. Instead of constructing new methodologies for different application scenarios, the methodology itself could include a step in which the engineers pick from a list of available methods, for example for requirements analysis, the ones suitable for a particular task and build up their own process model. Template process models covering standard requirements could be available.

Furthermore current methodologies are not integrated in a broader process model covering human, technological, and process aspects of knowledge management. These aspects are important to deploy knowledge management in a holistic manner. In this context the costs incurred by the ontology building effort become an issue. Estimating these costs is still vague businesses.

To summarize, the following issues are still open:

1. Ontology maintenance support.
2. Distributed ontology engineering.
3. Fine-grained guidelines for all phases.
4. Representation of multiple views.
5. Agreement support under conflicting interests.
6. Best practices.
7. Ontology engineering with the help of automated methods.

8. Process definition by single process step combination.
9. Integration into business process model.
10. Cost estimation and pricing.

This chapter has been able to survey only a few of the ontology engineering tools. Since the introduction of the plug-in concept to OEEs, the number of features available for the more established tools has increased tremendously and for many tasks one can find a tool supporting it. However, integration with different process models is still lacking. Some tools offer support for a specific process model, but none can be customized to provide guidance through an arbitrary process. As there are many tools offering different functionalities the slightly different implementations for the standardized representation languages hinders inter-operability. Besides these more procedurally oriented requirements, the technical solutions to support, for example versioning, ontology learning or distributed engineering of ontologies, are also in need of improvement.

9.4. DILIGENT METHODOLOGY

As a summary of the previous section we have identified a number of open research issues for ontology engineering. We will now present the current status of a new ontology engineering methodology. Due to space restrictions, we cannot introduce the complete methodology, but only the general process model. We then focus on a distinctive aspect of the process model in order to exemplify the level of detail of the methodology.

9.4.1. Process

The DILIGENT process model was conceived for knowledge sharing scenarios, which are dynamic and require the process model to cope with frequently changing user needs.[6]

The users: In DILIGENT there are several experts, with different and complementary skills, involved in collaboratively building the same ontology (see open issue 2). In the context of the Semantic Web and other noncentralized environments they may even belong to competing organizations and be geographically dispersed. Typically the domain experts involved in building the ontology are also its users. However, most ontology users will typically not build or modify the ontology. The

---

[6] In fact, we conjecture that the majority of knowledge sharing cases falls into this category. In particular in the Semantic Web context, we find these requirements.

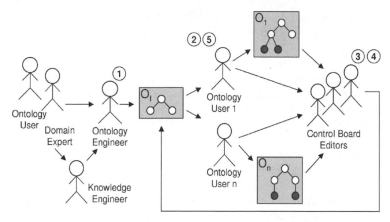

Figure 9.1 Roles and functions in distributed ontology engineering.

community of users is much larger than the community of active ontology builders.

Birds-eye view: An initial ontology is built by a small group. This ontology is made available and users are free to use and modify it locally for their own purposes. There is a central board that maintains and assures the quality of the shared core ontology. This central board is also responsible for updating the core ontology (see open issue 1). However, updates are mostly based on changes reoccurring at and requests by *locally* working users. Therefore, the board only *loosely controls* the process. Due to the changes introduced by the users over time that entail the changes introduced by the board, the ontology *evolves*.

Let us now survey the DILIGENT process at a finer level of granularity. DILIGENT comprises five main steps: (1) build, (2) local adaptation, (3) analysis, (4) revision, (5) local update (see Figure 9.1).

Build: The process starts by having *domain experts, users, knowledge engineers*, and *ontology engineers build* an initial ontology. In contrast to existing ontology engineering methodologies (cf. Section 9.3.1), we do not require completeness of the initial shared ontology with respect to the domain. The team involved in building the initial ontology should be relatively small, in order to easily find a small and consensual first version of the shared ontology.

Local adaptation: Once the core ontology is available, users work with it and, in particular, adapt it locally to their own needs. Typically, they will have their own business requirements and correspondingly change and adapt their local copy of the shared ontology (Noy and Klein, 2003; Stojanovic *et al.*, 2002). In their local environment, they are free to change the shared ontology. However, they are not allowed to directly change the shared ontology. The control board collects change requests to the shared ontology and logs local adaptations (either continuously or at control points).

Analysis: The board *analyses* the local ontologies and the requests for changes and tries to identify similarities in users' ontologies. We expect tools that allow an efficient analysis of the change requests. Since not all of the changes introduced or requested by the users will be introduced to the shared core ontology,[7] a crucial activity of the board is deciding which changes are going to be introduced in the next version of the shared ontology. The input from users provides the necessary arguments to underline change requests. A balanced decision that takes into account the different needs of the users and meets their evolving requirements[8] has to be found. Moreover, this decision must follow good knowledge representation practices.

Revise: The board should regularly *revise* the shared ontology, so that local ontologies do not diverge too far from the shared ontology. The goal of the revision is to realign the ontology with the obvious user needs and thus gaining higher acceptance, 'sharedness' and less local differences. Therefore, the board should have a well balanced and representative participation of the different kinds of participants involved in the process: knowledge engineers, domain experts, ontology engineers and users. In this case, users are involved in ontology development, at least through their requests and reoccurring improvements and by evaluating it, mostly from an usability point of view (understanding, actual advantage of use). Domain experts are responsible for evaluating it from a domain point of view ('does it represent the domain' or 'does it contain factual errors?'). Knowledge engineers in the board are responsible for evaluating the ontology, mostly from a domain and technical point of view (efficiency, standards conformance, logical properties like satisfiability). Ontology engineers too take responsibility for the technical evaluation, including analyzing and balancing arguments. Another possible task for the controlling board, although not always a requirement, is to assure some compatibility with previous versions. Revision can be regarded as a kind of ontology development guided by a carefully balanced subset of evolving user-driven requirements. Ontology engineers are responsible for updating the ontology, based on the decisions of the board. Revision of the shared ontology entails its evolution.

Local update: Once a new version of the shared ontology is released, users can *update* their own *local* ontologies to better use the knowledge represented in the new version. Even if the differences are small, users may rather reuse the new terms instead of using their previously locally defined terms that correspond to the new terms represented in the new version.

---

[7] The idea in this kind of development is not to merge all user ontologies.
[8] This is actually one of the trends in modern software engineering methodologies (see Rational Unified Process).

9.4.2. Argumentation Support

The process introduced in the previous sections requires the interaction of all participants at different stages in order to repeatedly build a consensual shared ontology. The different participants in the process exchange arguments to convey their opinions and eventually agree on a new version of the shared ontology. In order to increase acceptance, we want to provide easily accessible rationales for the decisions of the board and create a sense of community in building and evolving the ontology. The exchange of arguments should be embedded into a general argumentation framework, which offers conflict resolution strategies, ensures an efficient argumentation process and defines relevant roles (see open issue 5).

We formulated the hypothesis that such an argumentation framework can facilitate the ontology engineering and evolution process and offer a more fine-grained guidance to achieve agreement (see open issue 3). We studied the arguments and practices used in the evolution of the taxonomy of living beings which seem to point in the same direction. In order to substantiate this notion, we pursued an experiment in a Computer Science Department.

We performed the experiment in two steps: in the first, participants were not constrained in any way; in the second, participants were asked to use the argumentation framework, that is they followed stricter rules to conduct their discussions. The task in both sessions was to build an ontology, which represents the knowledge available in the research group, could be used for internal knowledge management, and made the research area comprehensible for outsiders. Both experiments lasted for 1½ hours. Concepts were only added after argumentation and some consensus was achieved.

The participants met in a virtual chat room. A moderator was responsible to remind people to stay on the subject and to include the modeling decisions into the formal ontology which was visualized on a web page. For this experiment very few procedural and methodological restrictions were *a priori* imposed. Participants agreed only on a few concepts, while the discussion was very chaotic and hard to follow, but gave us a good idea for how to improve the setting.

By analyzing the discussion, we identified out of the 30 different argument types proposed by the Rhetorical Structure Theory (Mann and Thompson, 1987), RST, those which had the most impact on the creation of the ontology: elaboration, evaluation/justification, examples, counter examples, alternatives.

With respect to the experimental setup we identified the following problems:

- participants started too many discussion threads and lost the overview,

- the discussion proceeded too fast, hence not everybody could follow the argumentation,
- it was difficult for the moderator to intervene, and
- there was no explicit possibility to vote or make decisions.

Even in this setting—where participants shared a very similar background knowledge—the creation of a shared conceptualization *without any guidance* was almost impossible, or at least very time consuming. We concluded that a more controlled approach is needed with respect to the process and moderation.

In the second experiment participants were asked to extend the ontology built in the first round. In this phase the formalism to represent the ontology was fixed. The most general concepts were also initially proposed, to avoid philosophical discussions. For the second round only the arguments elaboration, examples, counter examples, alternatives, evaluation/justification were allowed.

The participants in the second experiment joined two virtual chat rooms. One was used for providing topics for discussion, hand raising, and voting. The other one served to exchange arguments. When the participants—the same as in the first experiment—wanted to discuss a certain topic, for example the introduction of a new concept, they had to introduce it in the first chat room. The topics to discuss were published on a web site, and were processed sequentially. Each topic could then be justified with the allowed arguments. Participants could provide arguments only after hand raising and waiting for their turn. The participants decided autonomously when a topic had been sufficiently discussed, called for a vote and thus decided how to model a certain aspect of the domain. The evolving ontology was again published on a web site. The moderator had the same tasks as in the first experiment, but was stricter in interpreting the rules. Whenever needed, the moderator called for an example of an argument to enforce the participants to express their wishes clearly.

As expected the discussion was more focused, due to the stricter procedural rules. Agreement was reached more quickly and a much wider consensus was reached. With the stack of topics which were to be discussed (not all due to time constraints), the focus of the group was kept.

The restricted set of arguments is easy to classify and thus the ontology engineer was able to build the ontology in a straightforward way. It is possible to explain to new attendees why a certain concept was introduced and modeled in such a way, by simply pointing to the archived focused discussion. It is even possible to state the argumentation line used to justify it. The participants truly shared the conceptualization and did understand it. In particular, in conflict situations, when opinions diverged, the restriction of arguments was helpful. In this way participants could either prove their view, or were convinced.

More details on the argumentation model of DILIGENT can be found in Tempich *et al.* (2005).

9.5. FIRST LESSONS LEARNED

The analysis of the arguments and process driving the evolution of the taxonomy of living beings showed a high resemblance to the 5-step DILIGENT process and its accompanying argumentation framework. In two case studies, the argumentation framework and the process model of DILIGENT have been tested.

A case study in the tourism domain helped us to generally better comprehend the use of ontologies in a distributed environment. All users viewed the ontology mainly as a classification hierarchy for their documents. The ontology helped them share their own local view on documents with other users. Thus finding documents became easier.

Currently, we doubt that our manual approach for analyzing local structures will scale to cases with many more users. Therefore, we look to tools to recognize similarities in user behavior. Furthermore, the local update will be a problem when changes happen more often. Last, but not least, we have so far only addressed the ontology creation task itself—we have not yet measured explicitly if users get better and faster responses with the help of DILIGENT-engineered ontologies. All this remains work to be done in the future.

Despite the technical challenges, the users provided very positive feedback when asked, pointing to the integration into their daily work-flow through the use of a tool built by us and embedded in their work environment, which could easily be used.

Our experiment in a computer science department has given strong indication—though not yet full-fledged evidence—that a restriction of possible arguments can enhance the ontology engineering effort in a distributed environment. The restricted set of arguments will allow for reasoning over the debates, and maximizes the efficiency of the debate by providing only those kinds of arguments which proved to be the strongest in the debate. In addition, the second experiment underlines the fact that appropriate social management procedures and tool support help to reach consensus in a smoother way.

The process could certainly be enhanced with better tool support. Besides the argumentation stack, a stack for householding possible alternatives would be helpful. Arguments, in particular elaboration, evaluation and justification, and alternatives, were discussed heavily. However, the lack of appropriate evaluation measures made it difficult, at some times, for the contradicting opinions to achieve an agreement. In that case, the argumentation should be focused on the evaluation criteria.

9.6. CONCLUSION AND NEXT STEPS

In the last couple of years, we have witnessed a change of focus in the area of ontologies and ontology-based information systems: while the application of ontologies was restricted for a long time to academia projects, in the last 10 years ontologies have become increasingly relevant for commercial applications as well. A first prerequisite for the successful introduction of ontologies in the latter setting is the availability of proved and tested Ontology Engineering methodologies, which break down the complexity of typical engineering processes and offer guidelines to monitor them. Although existing methodologies have already proven to fulfill these requirements for a number of application scenarios, open issues remain to be researched in order for the methodologies to be applicable more widely.

With the DILIGENT methodology, we tackle some of these open issues and propose a methodology which allows continuous improvement of the underlying ontology in decentralized settings. Moreover we offer more fine-grained support to enhance the agreement process with an argumentation framework. However, the methodology is still under development and will be further developed to cover more aspects of the ontology engineering process. For example, the integration of ontology learning methods to automate the ontology building process (see open issue 7) is already covered in general by DILIGENT. However, due to the quality of the results of current ontology learning methods, this initial proposal is not sufficient and a more fine-grained process model is under development. A more fine-grained support for the evaluation of ontologies is being integrated into the methodology. Criteria to identify proper ontology evaluation schemes and tools for a more automatic appliance of such evaluation techniques are being developed within the DILIGENT methodology. These take into account the whole range of evaluation methods from philosophical notions (Völker et al., 2005) to logical satisfiability (Gomez-Perez et al., 2003).

We are currently integrating the process model into a knowledge management business process (see open issue 9). Regarding the estimation of costs incurred by the ontology building process (see open issue 10) a parametric cost estimation model is under development, which will be applied for our methodology. In the course of several projects in which our methodology is being applied, we will capture experiences and describe best practices (see open issue 6).

REFERENCES

Abecker A, Bernardi A, Hinkelmann K, Kuehn O, Sintek M. 1998. Toward a technology for organizational memories. *IEEE Intelligent Systems* 13(3):40–48.

Arpirez JC, Corcho O, Fernandez-Lopez M, Gomez-Perez A. 2001. WebODE: A scalable workbench for ontological engineering. In *Proceedings of the First International Conference on Knowledge Capture (K-CAP) October 21–23, 2001, Victoria, B.C., Canada.*

Bernaras A, Laresgoiti I, Corera J. 1996. Building and reusing ontologies for electrical network applications. In *Proceedings of the European Conference on Artificial Intelligence (ECAI'96).*

Berners-Lee T, Hendler J, Lassila O. 2001. The semantic web. *Scientific American*, 2001(5). available at http://www.sciam.com/2001/0501issue/0501berners-lee.html.

Bonifacio M, *et al.* 2003. Peer-mediated distributed knowldege management. In van Elst L, Dignum V, Abecker A, (eds), *Proceedings of the AAAI Spring Symposium "Agent-Mediated Knowledge Management (AMKM-2003)"*, Lecture Notes in Artificial Intelligence (LNAI) 2926, Berlin: Springer.

Bozsak E, Ehrig M, Handschuh S, Hotho A, Maedche A, Motik B, Oberle D, Schmitz C, Staab S, Stojanovic L, Stojanovic N, Studer R, Stumme G, Sure Y, Tane J, Volz R, Zacharias V. 2002. KAON—Towards a large scale semantic web. In Bauknecht K, Tjoa AM, Quirchmayr G (eds), *Proceedings of the Third International Conference on E-Commerce and Web Technologies (EC-Web 2002)*, Vol. 2455 of *LNCS*, Aix-en-Provence, France: Springer, pp 304–313.

Cristani M, Cuel R. 2005. A survey on ontology creation methodologies. *International Journal on Semantic Web and Information System* 1(2):49–69.

Davenport TH. 1996. Some principles of knowledge management. Technical report, Graduate School of Business, University of Texas at Austin, Strategy and Business.

Davies J, Fensel D, van Harmelen F (eds). 2002. *On-To-Knowledge: Semantic Web enabled Knowledge Management.* John Wiley and Sons, Ltd.

Dieng R, Corby O, Giboin A, Ribiere M. 1999. Methods and tools for corporate knowledge management. *International Journal of Human-Computer Studies* 51(3):567–598.

Duineveld AJ, Stoter R, Weiden MR, Kenepa B, Benjamins VR. 2000. Wondertools? A comparative study of ontological engineering tools. *International Journal of Human-Computer Studies* 6(52):1111–1133.

Ehrig M, Haase P, van Harmelen F, Siebes R, Staab S, Stuckenschmidt H, Studer R, Tempich C. 2003. The SWAP data and metadata model for semantics-based peer-to-peer systems. In *Proceedings of MATES-2003. First German Conference on Multiagent Technologies*, LNAI, Erfurt, Germany: Springer.

Fernandez-Lopez M, Gomez-Perez A, Sierra JP, Sierra AP. 1999. Building a chemical ontology using Methontology and the Ontology Design Environment. *Intelligent Systems* 14(1).

Gangemi A, Pisanelli D, Steve G. 1998. Ontology integration: Experiences with medical terminologies. In *Formal Ontology in Information Systems*, Guarino N (ed.). IOS Press: Amsterdam. pp 163–178.

Gomez-Perez A. 1996. A framework to verify knowledge sharing technology. *Expert Systems with Application* 11(4):519–529.

Gomez-Perez A. 2004. Ontology evaluation. In *Handbook on Ontologies*, Volume 10 of *International Handbooks on Information Systems*, chapter 13. Staab S, Studer R (eds). Springer: pp 251–274.

Gomez-Perez A, Angele J, Fernandez-Lopez M, Christophides V, Stutt A, Sure Y, *et al.* (2002). A survey on ontology tools. OntoWeb deliverable 1.3, Universidad Politecnia de Madrid.

Gomez-Perez A, Fernandez-Lopez M, Corcho O. 2003. *Ontological Engineering*. Advanced Information and Knowlege Processing. Springer.

Guarino N, Welty C. 2002. Evaluating ontological decisions with OntoClean. *Communications of the ACM* 45(2):61–65.

Holsapple CW, Joshi KD. 2002. A collaborative approach to ontology design. *Communications of the ACM* 45(2):42–47.

Hou CJ, Noy NF, Musen M. 2002. *A Template-based Approach Toward Acquisition of Logical Sentences*. In Musen *et al.*, 2002, pp 77–89.

IEEE. 1990. IEEE standard glossary of software engineering terminology. IEEE Standard 610.12-1990, ISBN 1-55937-067-X.

IEEE. 1996. IEEE guide for developing of system requirements specifications. IEEE Standard 1233–1996.

Jarrar M, Meersman R. 2002. Formal ontology engineering in the DOGMA approach. In Meersmann *et al.*, 2002), pp 1238–1254.

Kotis K, Vouros G. 2003. Human centered ontology management with HCONE. In *ODS'03: Proceedings of the IJCAI-03 Workshop on Ontologies and Distributed Systems*, volume 71. CEUR-WS.org.

Kotis K, Vouros GA, Alonso JP. 2004. HCOME: Tool-supported methodology for collaboratively devising living ontologies. In *SWDB'04: Second International Workshop on Semantic Web and Databases 29-30 August 2004 Co-located with VLDB*. Springer-Verlag.

Kunz W, Rittel HWJ. 1970. Issues as elements of information systems. Working Paper 131, Institute of Urban and Regional Development, University of California, Berkeley, California.

Leger A, Akkermans H, Brown M, Bouladoux J-M, Dieng R, Ding Y, Gomez-Perez A, Handschuh S, Hegarty A, Persidis A, Studer R, Sure Y, Tamma V, Trousse B. 2002a. Successful scenarios for ontology-based applications. OntoWeb deliverable 2.1, France Telecom R&D.

Leger A, Bouillon Y, Bryan M, Dieng R, Ding Y, Fernandez-Lopez M, Gomez-Perez A, Ecoublet P, Persidis A, Sure Y. 2002b. Best practices and guidelines. OntoWeb deliverable 2.2, France Telecom R&D.

Mann WC, Thompson SA. 1987. Rhetorical structure theory: A theory of text organization. In *The Structure of Discourse*, Polanyi L (ed.). Ablex Publishing Corporation: Norwood, NJ.

Motik B, Maedche A, Volz R. 2002. A conceptual modeling approach for semantics–driven enterprise applications. In Meersman R, Tari Z, *et al.* (eds). 2002. *Proceedings of the Confederated International Conferences: On the Move to Meaningful Internet Systems (CoopIS, DOA, and ODBASE 2002)*, Vol. 2519 of *Lecture Notes in Computer Science (LNCS)*, University of California, Irvine, USA. Springer, pp 1082–1099.

Musen M, Neumann B, Studer R (eds). 2002. *Intelligent Information Processing*. Kluwer Academic Publishers: Boston, Dordrecht, London.

Nicola AD, Navigli R, Missikoff M. 2005. Building an eProcurement ontology with UPON methodology. In *Proceedings of 15th e-Challenges Conference*, Ljubljana, Slovenia.

Noy N, Fergerson R, Musen M. 2000. The knowledge model of *Protégé-2000*: Combining interoperability and flexibility. Vol. 1937 of *Lecture Notes in Artificial Intelligence (LNAI)*, Juan-les-Pins, France. Springer, pp 17–32.

Noy N, Klein M. 2003. Ontology evolution: Not the same as schema evolution. *Knowledge and Information Systems*.

Noy, N McGuinness D L. 2001. Ontology development 101: A guide to creating your first ontology. Technical Report KSL-01-05 and SMI-2001-0880, Stanford Knowledge Systems Laboratory and Stanford Medical Informatics.

Pinto HS, Martins J. 2001. A methodology for ontology integration. In *Proceedings of the First International Confrence on Knowledge Capture (K-CAP2001)*, New York. ACM Press, pp 131–138.

Schreiber G, Akkermans H, Anjewierden A, de Hoog R, Shadbolt N, van de Velde W, Wielinga B. 1999. *Knowledge Engineering and Management—The Common-KADS Methodology*. The MIT Press: Cambridge, Massachusetts; London, England.

Spyns P, Meersman R, Jarrar M. 2002. Data modelling versus ontology engineering. *SIGMOD Record—Web Edition*, 31(4). Special Section on Semantic Web and Data Management, Meersman R, Sheth A (eds). Available at http://www.acm.org/sigmod/record/.

Staab S, Schnurr H-P. 2002. Knowledge and business processes: Approaching an integration. In *Management and Organizational Memories, Dieng-Kuntz R, Matta N (eds)*. *Knowledge* Kluwer Academic Publishers: Boston, Dordrecht, London. pp 75–88.

Staab S, Studer R (eds). 2004. *Handbook on Ontologies in Information Systems*. International Handbooks on Information Systems. Springer.

Stojanovic L, Maedche A, Motik B, Stojanovic N. 2002. User-driven ontology evolution management. In *Proceedings of the 13th European Conference on Knowledge Engineering and Knowledge Management EKAW*, Madrid, Spain.

Sure Y. 2003. *Methodology, Tools and Case Studies for Ontology based Knowledge Management*. PhD thesis, University of Karlsruhe.

Sure Y, Angele J (eds). 2002. *Proceedings of the First International Workshop on Evaluation of Ontology based Tools (EON 2002)*, Vol. 62 of *CEUR Workshop Proceedings*, Siguenza, Spain. CEUR-WS Publication, available at http://CEUR-WS.org/Vol-62/.

Sure Y, Angele J, Staab S. 2003. OntoEdit: Multifaceted inferencing for ontology engineering. *Journal on Data Semantics*, LNCS(2800):128–152.

Sure Y, Erdmann M, Angele J, Staab S, Studer R, Wenke D. 2002. OntoEdit: Collaborative ontology development for the semantic web. In Horrocks I, Hendler JA (eds). *Proceedings of the First International Semantic Web Conference: The Semantic Web (ISWC 2002)*, volume 2342 of *Lecture Notes in Computer Science (LNCS)*, pp 221–235. Sardinia, Italy. Springer.

Swartout B, Ramesh P, Knight K, Russ T. 1997. Toward distributed use of largescale ontologies. In *Symposium on Ontological Engineering of AAAI*, Stanford, CA.

Tempich C, Pinto HS, Sure Y, Staab S. 2005. An argumentation ontology for distributed, loosely-controlled and evolving engineering processes of ontologies (DILIGENT). In Bussler C, Davies J, Fensel D, Studer R (eds). *Second European Semantic Web Conference, ESWC 2005*, LNCS, Heraklion, Crete, Greece. Springer.

Tempich C, Volz R. 2003. Towards a benchmark for semantic web reasoners—ananalysis of the DAML ontology library. In Sure YM (ed.). *Evaluation of*

Ontology-based Tools (EON2003) at Second International Semantic Web Conference (ISWC 2003).

Uschold M, Grueninger M. 1996. Ontologies: principles, methods and applications. *Knowledge Sharing and Review* 11(2).

Uschold M, King M. 1995. Towards a methodology for building ontologies. In *Workshop on Basic Ontological Issues in Knowledge Sharing, held in conjunction with IJCAI-95*, Montreal, Canada.

Uschold M, King M, Moralee S, Zorgios Y. 1998. The enterprise ontology. *Knowledge Engineering Review* 13(1):31–89.

Völker J, Vrandecic D, Sure Y. 2005. Automatic evaluation of ontologies (AEON). In *Proceedings of the Fourth International Semantic Web Conference (ISWC'05)*, Galway, Ireland.

10

Semantic Web Services – Approaches and Perspectives

Dumitru Roman, Jos de Bruijn, Adrian Mocan, Ioan Toma,
Holger Lausen, Jacek Kopecky, Christoph Bussler, Dieter Fensel,
John Domingue, Stefania Galizia and Liliana Cabral

10.1. SEMANTIC WEB SERVICES – A SHORT OVERVIEW

Web services (Alonso *et al.*, 2001) – pieces of functionalities which are accessible over the Web – have added a new level of functionality to the current Web by taking a first step towards seamless integration of distributed software components using Web standards. Nevertheless, current Web service technologies around SOAP (XML Protocol Working Group, 2003), WSDL (WSDL, 2005), and UDDI (UDDI, 2004) operate at a syntactic level and, therefore, although they support interoperability (i.e., interoperability between the many diverse application development platforms that exist today) through common standards, they still require human interaction to a large extent: the human programmer has to manually search for appropriate Web services in order to combine them in a useful manner, which limits scalability and greatly curtails the added economic value of envisioned with the advent of Web services (Fensel and Bussler, 2002). For automation of tasks, such as Web service discovery, composition and execution, semantic description of Web services is required (McIlraith *et al.*, 2001).

Recent research aimed at making Web content more machine processable, usually subsumed under the common term Semantic Web (Berners-Lee *et al.*, 2001) are gaining momentum also, in particular, in the context of Web services usage. Here, semantic markup shall be exploited to automate the

Semantic Web Technologies: Trends and Research in Ontology-based Systems
John Davies, Rudi Studer, Paul Warren © 2006 John Wiley & Sons, Ltd

tasks of Web service discovery, composition, and invocation, thus enabling seamless interoperation between them while keeping human intervention to a minimum. The description of Web services in a machine-understandable fashion is expected to have a great impact in areas of e-Commerce and Enterprise Application Integration, as it is expected to enable dynamic and scalable cooperation between different systems and organizations: Web services provided by cooperating businesses or applications can be automatically located based on another business or application needs, they can be composed to achieve more complex, added-value functionalities, and cooperating businesses or applications can interoperate without prior agreements custom codes. Therefore, much more flexible and cost-effective integration can be achieved.

In order to provide the basis for Semantic Web Services, a fully fledged framework needs to be provided: starting with a conceptual model, continuing with a formal language to provides formal syntax and semantics (based on different logics in order to provide different levels of logical expressiveness) for the conceptual model, and ending with an execution environment, that glue all the components that use the language for performing various tasks that would eventually enable automation of service. In this context, this chapter gives an overview of existing approaches to Semantic Web Services and highlights their features as far as such a fully fledged framework for SWS is concerned. We start by introducing, in Section 10.2, the most important European initiative in the area of SWS – the WSMO approach to SWS. In Section 10.3, we provide an overview of OWL-S – an OWL-based Web service ontology, and in Section 10.4 the SWSF – a language and an ontology for describing services. Furthermore, we look also at other approaches – IRS III (in Section 10.5) and WSDL-S (in Section 10.6) – that although do not aim at providing a fully fledged framework for SWS, tackle some relevant aspects of SWS. In Section 10.7, we take a closer look at the gap – usually called 'grounding' – between the semantic and the syntactic descriptions of services, and identify several approaches to deal with the grounding in the context of SWS. Section 10.8 concludes this chapter and points out perspectives for future research in the area of Semantic Web Services.

10.2. THE WSMO APPROACH

The WSMO initiative[1], part of the SDK Cluster[2], is the major initiative in the area of SWS in Europe and has the aim of standardizing a unifying framework for SWS which provides support for conceptual modeling and formally representing services, as well as for automatic execution of services. In this Section we provide a general overview of the elements

---

[1]http://www.wsmo.org
[2]http://www.sdk-cluster.org/

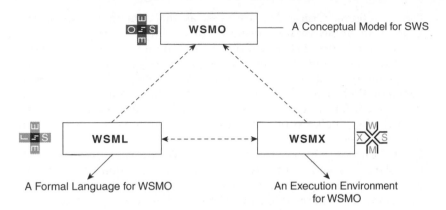

Figure 10.1 The WSMO approach to SWS.

that are part of the WSMO approach to SWS (see Figure 10.1): the Web service Modeling Ontology (WSMO) – a conceptual model for Semantic Web Services (Section 10.2.1), the Web Service Modeling Language (WSML) – a language which provides a formal syntax and semantics for WSMO (Section 10.2.2), and the Web Service Modeling Execution Environment (WSMX) – an execution environment, which is a reference implementation for WSMO, offering support for interacting with Semantic Web Services (Section 10.2.3).

10.2.1. The Conceptual Model – The Web Services Modeling Ontology (WSMO)

WSMO (Roman *et al.*, 2005) provides ontological specifications for the core elements of Semantic Web services. In fact, Semantic Web services aim at an integrated technology for the next generation of the Web by combining Semantic Web technologies and Web services, thereby turning the Internet from a information repository for human consumption into a world-wide system for distributed Web computing. Therefore, appropriate frameworks for Semantic Web services need to integrate the basic Web design principles, those defined for the Semantic Web, as well as design principles for distributed, service-orientated computing of the Web. WSMO is, therefore, based on the following design principles:

- *Web Compliance*: WSMO inherits the concept of Universal Resource Identifier (URI) for unique identification of resources as the essential design principle of the Word-Wide Web. Moreover, WSMO adopts the concept of Namespaces for denoting consistent information spaces, supports XML and other W3C Web technology recommendations, as well as the decentralization of resources.

- *Ontology Based*: Ontologies are used as the data model throughout WSMO, meaning that all resource descriptions as well as all data interchanged during service usage are based on ontologies. Ontologies are a widely accepted state-of-the-art knowledge representation, and have thus been identified as the central enabling technology for the Semantic Web. The extensive usage of ontologies allows semantically enhanced information processing as well as support for interoperability; WSMO also supports the ontology languages defined for the Semantic Web.
- *Strict Decoupling*: Decoupling denotes that WSMO resources are defined in isolation, meaning that each resource is specified independently without regard to possible usage or interactions with other resources. This complies with the open and distributed nature of the Web.
- *Centrality of Mediation*: As a complementary design principle to strict decoupling, mediation addresses the handling of heterogeneities that naturally arise in open environments. Heterogeneity can occur in terms of data, underlying ontology, protocol, or process. WSMO recognizes the importance of mediation for the successful deployment of Web services by making mediation a first class component of the framework.
- *Ontological Role Separation*: Users, or more generally clients, exist in specific contexts which will not be the same as for available Web services. For example, a user may wish to book a holiday according to preferences for weather, culture, and childcare, whereas Web services will typically cover airline travel and hotel availability. The underlying epistemology of WSMO differentiates between the desires of users or clients and available services.
- *Description versus Implementation*: WSMO differentiates between the descriptions of Semantic Web services elements (description) and executable technologies (implementation). While the former requires a concise and sound description framework based on appropriate formalisms in order to provide a concise for semantic descriptions, the latter is concerned with the support of existing and emerging execution technologies for the Semantic Web and Web services. WSMO aims at providing an appropriate ontological description model, and to be complaint with existing and emerging technologies.
- *Execution Semantics*: In order to verify the WSMO specification, the formal execution semantics of reference implementations like WSMX as well as other WSMO-enabled systems provide the technical realization of WSMO. This principle serves as a mean to precisely define the functionality and behavior of the systems that are WSMO compliant.
- *Service versus Web service*: A Web service is a computational entity which is able to achieve a user goal by invocation. A service, in contrast, is the actual value provided by this invocation (Baida, 2005;

Preist, 2004)[3]. WSMO provides means to describe Web services that provide access (searching, buying, etc.) to services. WSMO is designed as a means to describe the former and not to replace the functionality of the latter.

The following briefly outlines the conceptual model of WSMO. The elements of the WSMO ontology are defined in a meta-meta-model language based on the Meta Object Facility (MOF) (OMG, 2002). MOF defines an abstract language and framework for specifying, constructing, and managing technology neutral meta-models. Since WSMO is meant to be a meta-model for Semantic Web Services, MOF was identified as the most suitable language/framework for defining the WSMO elements. In terms of the four MOF layers (meta-meta-model, meta model, model layer, and information layer), the language defining WSMO corresponds to the meta-meta-model layer, WSMO itself constitutes the meta-model layer, the actual ontologies, Web services, goals, and mediators specifications constitute the model layer, and the actual data described by the ontologies and exchanged between Web services constitute the information layer (the information layer in this context is actually related to the notion of grounding in the context of SWS, and which will be discussed in Section 10.7). The most frequently used MOF meta-modeling construct for the definition of WSMO elements is the **Class** construct (and implicitly its class-generalization **sub-Class** construct), together with its **Attributes**, the **type** of the Attributes, and their multiplicity specifications[4].

In order to allow complete item descriptions, every WSMO element is described by nonfunctional properties. These are based on the Dublin Core (DC) Metadata Set (Weibel *et al.*, 1998) for generic information item descriptions, and other service-specific properties related to the quality of service[5].

Ontologies: It provide the formal semantics for the terminology used within all other WSMO components. Using MOF, we define an ontology as described in the listing below:

```
Class ontology
   hasNonFunctionalProperties type nonFunctionalProperties
   importsOntology type ontology
   usesMediator type ooMediator
```

---

[3]Note that (Preist, 2004) also distinguishes between a computational entity in general and Web service, where the former does not necessarily have a Web accessible interface. WSMO does not make this distinction.

[4]Note that, for readability purposes, we avoid the usage of 'Attribute' keyword in the listings in which we define the WSMO top-level elements; the attributes of a Class (i.e., of a WSMO element) are defined on separate lines inside each listing.

[5]For a detailed description of all the elements defined in WSMO, we refer the reader to Roman *et al.* (2005).

```
hasConcept type concept
hasRelation type relation
hasFunction type function
hasInstance type instance
hasAxiom type axiom
```

A set of *non-functional properties* are available for characterizing ontologies; they usually include the DC Metadata elements. *Imported ontologies* allow a modular approach for ontology design and can be used as long as no conflicts need to be resolved between the ontologies. When importing ontologies in realistic scenarios, some steps for aligning, merging, and transforming imported ontologies in order to resolve ontology mismatches are needed. For this reason *ontology mediators* are used (*OO Mediators*). *Concepts* constitute the basic elements of the agreed terminology for some problem domain. *Relations* are used in order to model interdependencies between several concepts (respectively instances of these concepts); *functions* are special relations, with a unary range and a n-ary domain (parameters inherited from relation), where the range value is functionally dependent on the domain values, and *instances* are either defined explicitly or by a link to an instance store, that is, an external storage of instances and their values.

Web services: WSMO provides service descriptions for describing services that are requested by service requesters, provided by service providers, and agreed between service providers and requesters. In the listing below, the common elements of these descriptions are presented.

```
Class webService
  hasNonFunctionalProperties type nonFunctionalProperties
  importsOntology type ontology
  usesMediator type {ooMediator, wwMediator}
  hasCapability type capability multiplicity = single-valued
  hasInterface type interface
```

Within the service class the *non-functional properties* and *imported ontologies* attributes play a role that is similar to that found in the ontology class with the minor addition of a quality of service nonfunctional property. An extra type of *mediator* (*WW Mediator*) is also included, in order to deal with protocol and process-related mismatches between Web services.

The final two attributes define the two core WSMO notions for semantically describing Web services: a *capability* which is a functional description of a Web Service, describing constraints on the input and output of a service through the notions of preconditions, assumptions, postconditions, and effects; and Web service *interfaces* which specify how the service behaves in order to achieve its functionality. A service interface consists of a *choreography* which describes the interface for the client-service interaction required for service consumption, and an *orchestration* which describes how the functionality of a Web service is achieved by aggregating other Web services.

Goals: A goal specifies the objectives that a client may have when consulting a web service, describing aspects related to user desires with respect to the requested functionality and behavior. Ontologies are used as the semantically defined terminology for goal specification. Goals model the user view in the Web Service usage process and therefore are a separate top level entity in WSMO.

```
Class goal
  hasNonFunctionalProperties type nonFunctionalProperties
  importsOntology type ontology
  usesMediator type {ooMediator, ggMediator}
  requestsCapability type capability multiplicity = single-valued
  requestsInterface type interface
```

As presented in listing above, the *requested capability* in the definition of a goal represents the functionality of the services the user would like to have, and the *requested interface* represents the interface of the service the user would like to have and interact with.

Mediators: The concept of Mediation in WSMO addresses the handling of heterogeneities occurring between elements that shall interoperate by resolving mismatches between different used terminologies (data level), on communicative behavior between services (protocol level), and on the business process level. A WSMO Mediator connects the WSMO elements in a loosely-coupled manner, and provides mediation facilities for resolving mismatches that might arise in the process of connecting different elements defined by WSMO. The description elements of a WSMO Mediator are its source and target elements, and the mediation service for resolving mismatches, as shown in the listing below.

```
Class mediator
  hasNonFunctionalProperties type nonFunctionalProperties
  importsOntology type ontology
  hasSource type {ontology, goal, webService, mediator}
  hasTarget type {ontology, goal, webService, mediator}
  hasMediationService type {goal, webService, wwMediator}
```

WSMO defines different types of mediators for connecting the distinct WSMO elements: *OO Mediators* connect and mediate heterogeneous ontologies, *GG Mediators* connect Goals, *WG Mediators* link Web services to Goals, and *WW Mediators* connects interoperating Web services resolving mismatches between them.

10.2.2. The Language – The Web Service Modeling Language (WSML)

The WSML (de Brujin, 2005) is a language for the description of ontologies, goals, Web services, and mediators based on the conceptual model of WSMO. WSML provides one coherent framework which brings together Web technologies with different well-known logical language paradigms. We take Description Logics (Baader, 2003), Logic Programming (Lloyd, 1987), and F-Logic (Kifer *et al.* 1995), as starting points for the development of a number of WSML language variants. WSML can be seen as a testing ground for the development of formal techniques for Web Service description. In order to have full freedom in development of the language, we have chosen to initially develop WSML independently from existing Semantic Web and Web Service standards[6]. There are ongoing efforts to relate WSML to existing standards. With WSML we take a top-down approach to Semantic Web Service description. So far, the focus has been on the use of different formalisms for describing static knowledge (ontologies) related to the Web services. There are ongoing efforts to investigate the use of formal methods to describe the dynamics of services.

There currently exists no language for the description of Semantic Web Services which takes into account all aspects identified by WSMO; a language is required for the description of WSMO services, in order to take into account all aspects of Web service modeling identified in WSMO. Other approaches (e.g., OWL-S (2004), are based on existing languages which are constructed for different purposes (e.g., OWL (Dean and Schreiber, 2004)); it is not clear whether these languages are the appropriate languages for the description of Semantic Web Services. WSML takes different formalisms in order to investigate their applicability for the description of Semantic Web Services. Since our goal is to investigate the applicability of different formalisms to the description of Semantic Web Services, it would be too restrictive to base our effort on an existing language recommendation. A major goal in our development of WSML is to investigate the applicability of different formalisms, most notably Description Logics and Logic Programming, in the area of Web services. Furthermore, a future goal of WSML is to investigate the combination of static and dynamic knowledge of services.

We see three main areas which benefit from the use of formal methods in service description:

---

[6] Note that WSML takes into account the concepts of URI and IRI, and relates to XML and RDF, but was not developed with these in mind.

- Ontology description
- Declarative functional description of Goals and Web services
- Description of dynamics

In its current version WSML defines a syntax and semantics for ontology description. The formalisms which were mentioned earlier are used to give a formal meaning to ontology descriptions in WSML. For the functional description of Goals and Web services, WSML offers a syntactical framework, with Hoare-style semantics in mind (Hoare, 1969). However, WSML does not formally specify the semantics of the functional description of services. A possible direction for this semantics description is the use of Transaction Logic (e.g., like in Kifer *et al.* (2004)). The description of the dynamics of Web services (choreography and orchestration) in the context of WSML is currently under investigation, but has not (yet) been integrated in WSML.

This section is further structured as follows. We first motivate and describe the formalisms which form the basis for WSML, as well as the WSML language variants which are based on these formalisms, after which we briefly introduce the syntax and semantics of WSML, taking as example a simplified version of the Amazon Web Service.

10.2.2.1. WSML Language Variants

WSML has language variants which are based on five formalisms related to First-Order Predicate Logic (FOPL). Description Logics (DL) (Baader, 2003) is a family of languages which (for the most part) can be seen as subsets of FOPL. We have chosen to use the DL language SHIQ in WSML because it is an expressive DL for which efficient sound and complete reasoning algorithms and implementations exist for checking concept satisfiability, subsumption, and other reasoning tasks. Furthermore, there exist application in Web Service discovery and the language has already been applied to the Semantic Web in the language OWL (Dean and Schreiber, 2004).

Another formal pillar of WSML is Logic Programming[7]. Logic Programming has a wide body of research work in the area of query answering, as well as many efficient implementations. Furthermore, there exist applications of Logic Programming in the area of Web Service for discovery, contracting, and other tasks and there is also a broad interest in applying rule languages to the Web (http://www.ruleml.org/) better cite RIF working group.

F-Logic (Kifer *et al.*, 1995) is an extension of FOPL with higher-order style Object Oriented modeling primitives which stays semantically in a First-Order framework. With F-Logic Programming we mean the Logic Programming language which is obtained from the Horn subset of F-Logic.

---

[7]When talking about Logic Programming we mean purely declarative rules languages, based on the so-called Horn subset of FOPL, with a model-theoretic semantics based on Herbrand models.

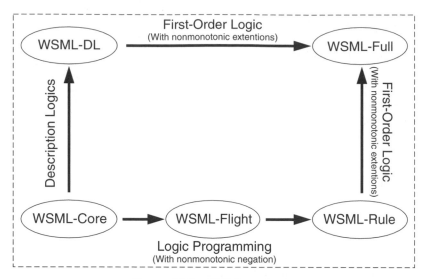

Figure 10.2. WSML variants.

The WSML syntax inspired by F-Logic arguably makes logical expressions easier to write since the modeling vocabulary is not restricted to predicates, as in FOPL, but also includes concepts, instances, attribute typing, and attribute values. There exist several implementations of F-Logic Programming as well as experiences in case studies, as well as a commercial product (Ontobroker). F-Logic Programming can be reduced to regular Logic Programming, thereby benefiting from the research and experience in the area.

Figure 10.2 depicts the WSML language variants. WSML-Core marks the intersection of Description Logics and Logic Programming, also called Description Logic Programs (Grosof *et al.*, 2003). WSML extends this core language in two directions, namely DLs and F-Logic Programming, which allows the variants to benefit from the established research and the tools which have been developed in these areas. WSML-DL is based on the DL SHIQ(D); WSML-Flight is based on the Datalog subset of F-Logic and WSML-Rule is based on the Horn subset of F-Logic; WSML-Flight and WSML-Rule both include negation-as-failure. WSML-Full unifies the Description Logic and Logic Programming paradigms under a First-Order umbrella with specific extensions to capture the negation-as-failure of the Logic Programming-based WSML variants.

In WSML we use a subset of F-Logic as a syntactic extension of the variant based on Logic Programming, namely WSML-Rule. Interoperability between the variants based on DL and Logic Programming can be achieved using so-called Description Logic Programs (DLP) (Grosof *et al.* 2003). DLP prescribes syntactical restrictions on the ontologies such that they both can be seen as DL ontologies and Logic Programs (LP). The relationship between the syntactically equivalent DL ontology and the LP is then as follows: the DL ontology and the LP are equivalent with respect to entailment of ground atomic formulae. Thus, the extension of WSML-Core to WSML-Flight is a syntactic extension, but also a semantic restriction, since WSML-Flight does not define entailment of nonground formulae.

10.2.2.2. WSML Syntax

WSML provides three syntaxes for the description of Semantic Web Services, based on WSMO. WSML has a surface syntax, as well as XML and RDF syntaxes for exchange over the Web. WSML can be seen as a testing ground for using formal methods in the description of Semantic Web Services.

In the following we will briefly outline the main features of the WSML surface syntax using an example from the book buying domain. Because of space limitations, we will not discuss the XML and RDF syntaxes. The example consists of a small ontologies which describes books, as well as a small Web Service description which describes the functionality of adding a book to a shopping cart, given the id of the cart and the item to be added, which is an actual operation offered by the Amazon Web Service.

```
wsmlVariant_"http://www.wsmo.org/wsml/wsml-syntax/wsml-flight"
namespace {_"http://example.org/amazonOntology#",
  dc _"http://purl.org/dc/elements/1.1/"}
ontology_"http://example.org/amazonOntology"
  concept amazonBook
    title ofType_string
  concept cart
    id ofType (1)_string
    items ofType amazonBook
webService_"http://example.org/amazonService"
  importsOntology_"http://example.org/amazonOntology"
  capability
    sharedVariables {?cartId, ?item}
    precondition
      definedBy
        ?cartId memberOf_string and ?item memberOf amazonBook.
    postcondition
      definedBy
        forall ?cart (?cart[id hasValue ?cartId] memberOf cart implies
              ?cart[items hasValue ?item]).
```

The ontology ('amazonOntology') describes books and shopping carts at Amazon. The description of the Web Service shows how concepts from the ontology are used in the Web Service description. The precondition in the example requires that there is an input of type string (concepts preceded with an underscore '_' are built-in datatypes, corresponding to the XML Schema datatypes) and an input of type amazonBook. The postcondition describes that if there is indeed a cart with the particular id, the book will be in the cart.

WSML has a so-called 'conceptual syntax' which reflects the conceptual model of WSMO in the modeling of Ontologies, Web services, Goals, and Mediators; in the example, all statements belong to the conceptual syntax, except for the logical expressions under the precondition and the postcondition, respectively. The use of the conceptual syntax of WSML does not depend on the chosen WSML variant, except for a few restrictions on attribute constraints in the conceptual syntax for WSML-Core and WSML-DL, which are a result of the lack of nonmonotonic negation in these variants.

Besides the conceptual syntax, WSML has a logical syntax which reflects the underlying formalism of the particular WSML variant; the logical expressions under the keyword *definedBy* in the example are part of the logical syntax. The syntax is based on FOPL syntax where the logical connectives are English language words ('and,' 'or,' etc.) and with specific extensions for frame-based modeling inspired by F-Logic (e.g., the keyword *memberOf* in the example). Each WSML variant defines restrictions on the logical syntax. For example, nonmonotonic negation ('naf') may not be used in WSML-Core and WSML-DL, and existential quantification ('exists') may not be used in WSML-Core, WSML-Flight, and WSML-Rule.

The separation between conceptual and logical modeling allows for an easy adoption by nonexperts, since the conceptual syntax does not require expert knowledge in logical modeling, whereas complex logical expressions require more familiarity and training with the language. Thus, WSML allows the modeling of different aspects related to Web services on a conceptual level, while still offering the full expressive power of the logic underlying the chosen WSML variant. The conceptual syntax for ontologies has an equivalent in the logical syntax. This correspondence is used to define the semantics of the conceptual syntax. The translation between the conceptual and logical syntax is sketched in Table 10.1; we

Table 10.1 Translating conceptual to logical syntax.

Conceptual	Logical
concept A subConcepOf B	A subConceptOf B

concept A
  B ofType (0 1) C
	A[B ofType C]
	! – ?x memberOf A and
	   ?x[B hasValue ?y, B hasValue ?z] and ?y != ?z

concept A B ofType C relation A/n subRelationOf B	A[B ofType C]. A(x 1,..., x n) implies B(x 1,..., x n)
instance A memberOf B C hasValue D	A memberOf B. A[C hasValue D].

refer the reader to de Brujin (2005) for a complete translation of all the constructs of the conceptual syntax to the constructs of the logical syntax.

10.2.2.3. WSML Semantics

The semantics of WSML ontologies is defined through a mapping between the logical syntax and the formalism which underlies the variant. WSML-Core and WSML-DL logical expressions are mapped to FOPL. The restrictions on the WSML syntax for these variants ensures that the expressions stay inside the expressiveness of the SHIQ description logic. The semantics of WSML-Flight and WSML-Rule is defined through a mapping to F-Logic Programming (Kifer *et al.*, 1995, Appendix A). Space limitations prevent us from given the complete semantics of WSML. Instead, we give a rough sketch of the WSML semantics and refer the interested reader to de Brujin (2005).

In order to facilitate the mapping between the WSML variant and the underlying logic, the WSML ontology is first transformed to a set of normalized logical expressions, according to Table 10.1. Then, this normalized set of logical expressions is mapped to the logic using the mapping function π. Finally, satisfiability and entailment are defined w.r.t. this transformed WSML ontology.

The semantics of WSML is defined through a mapping to a well-known formalism rather than giving a direct semantics, in order to facilitate understanding of the language and so that complexity results follow immediately from the definitions and the known literature. In order to give an idea of the semantics, we show the example ontology amazonBooks transformed to F-Logic Programming:

$$\leftarrow x[y \Rrightarrow z] \wedge w[y \twoheadrightarrow v] \wedge \mathrm{not}\ v : z$$
$$amazon\,Book[title \Rrightarrow string]$$
$$cart[id \Rrightarrow string]$$
$$\leftarrow x : cart \wedge \mathrm{not}\,x[id \twoheadrightarrow y]$$
$$\leftarrow x : cart \wedge x[id \twoheadrightarrow y] \wedge x[id \twoheadrightarrow z] \wedge y \neq z$$
$$cart[items \Rrightarrow amazon\,Book]$$

Note that in the example, a rule without a head is an integrity constraint. A model of the rule base violates an integrity constraint if the body of the constraint is true in the model. We call a transformed WSML-Flight ontology O satisfiable iff. the transformed rule base $\pi(O)$ (without the constraints) has a perfect model M_O (Kifer *et al.*, 1995, Appendix A) and this model does not violate any of the integrity constraints. Furthermore, a satisfiable WSML-Flight ontology O entails a ground formula F iff. $\pi(F)$ is true in M_O.

10.2.3. The Execution Environment – The Web Service Modeling Execution Environment (WSMX)

Web Service Execution Environment (WSMX) is an execution environment which enables discovery, selection, mediation, and invocation of Semantic Web Services (Cimpian *et al.*, 2005). WSMX is based on the conceptual model provided by WSMO, being at the same time a reference implementation of it. It is the scope of WSMX to provide a testbed for WSMO and to prove its viability as a mean to achieve dynamic interaperatability of Semantic Web Services. In this section we briefly present the WSMX functionality (in Section 10.2.3.1) and its external behavior (in Section 10.2.3.2), followed by a short overview of the WSMX architecture (in Sections 10.2.3.2 and 10.2.3.3).

10.2.3.1. WSMX Functionality

WSMX functionality can be split in two main parts: first is the functionality that should be part of any environment for Semantic Web Services and second, the additional functionality coming from the enterprise system features of the framework. In the first case, the overall WSMX functionality can be seen as an aggregation of the components' functionalities, which are part of the WSMX architecture. In the second case WSMX offers features such as a plug-in mechanism that allows the integration of various distributed components, an internal workflow engine capable of executing formal descriptions of the components behavior or a resource manager that enables the persistency of all WSMO and nonWSMO data produced during run time.

The main components that have been already designed and implemented in WSMX, as described in (Cimpian *et al.* 2005) are the Core Component, Resource Manager, Discovery, Selection, Data and Process Mediator, Communication Manager, Choreography Engine, Web Service Modeling Toolkit, and the Reasoner.

The *Core Component* is the central component of the system connecting all the other components and managing the business logic of the system.

The *Resource Manager* manages the set of repositories responsible for the persistency of all the WSMO and non-WSMO related data flowing through the system. It is offering an uniform and in the same time the only (in the framework) point of access to potentially heterogeneous implementation of such repositories.

The *Discovery* component has the role of locating the services that fulfill a specific user request. This task is based on the WSMO conceptual framework for discovery (Keller *et al.*, 2004) which envision three main steps in this process: Goal Discovery, Web Service Discovery, and Service Discovery. The WSMX Service Discovery currently covers only the matching of a user's goal against different service descriptions based on syntactical consideration.

In case that more than one suitable service are found WSMX offers support for choosing only one of them; this operation is performed by the Selection component by applying different techniques ranging from simple 'always the first' to multi-criteria selection of variants (e.g., Web services nonfunctional properties as reliability, security, etc.) and interactions with the requester.

Two types of mediators are provided by WSMX to resolve the heterogeneity problems on data and process level. The *Data mediation* is based on paradigms of ontologies, ontologies mappings, and alignment with direct application on instance transformation. The *Process mediation* offers functionality for runtime analysis of two given patterns (i.e., WSMO choreographies) and compensates the possible mismatches that may appear.

The *Communication Manager* through its two subcomponents, the Receiver and the Invoker, enables the communication between the requester and the provider of the services. The invocation of services is based on the underlying communication protocol used by the service provider, and it is the responsibility of an adapter framework to implement the interactions that require different communication protocols than SOAP.

The *Choreography* engine provides means to store and retrieve choreography interface definitions, initiates the communication between the requester and the provider in direct correlation with the results returned by the Process Mediator, and keeps track of the communication state on both the provider and the requester sides. In addition it provides grounding information to the communication manager to enable any ordinary Web Service invocation.

Even if the *Reasoner* is not part of the WSMX development effort, a WSML compliant reasoner is required by various components as Data Mediator, Process Mediator, and Discovery.

The *Web Services Modeling Toolkit (WSMT)* is a collection of tools for Semantic Web services developed for the Eclipse framework. An initial set of tools exist including the WSMO Visualizer for viewing and editing WSML documents using directed graphs, a WSMX Management tool for managing and monitoring the WSMX environment, and a WSMX Data Mediation tool for creating mappings between WSMO ontologies.

10.2.3.2. WSMX External Behavior

WSMX external behavior is described in terms of the so-called entry points which represent standard interfaces that enable communication with external entities. There are four mandatory entry points that have to be available in each working instance of the system. Each of these entry points triggers a particular execution semantic[8], which on its turn selects the set of components to be used for that particular scenario. More details about WSMX Execution Semantics can be found in (Zaremba and Oren, 2005). As described in (Zaremba *et al.*, 2005) the four possible execution semantics are:

- *One-way Goal Execution*: This entry point allows the realization of a goal without any back and forth interactions. In this simplistic scenario the requester has to provide a formal description of its goal in WSML, and the data required for the invocation and the system will select and execute the service on behalf of the requester. The requester might receive a final confirmation, but this step is optional.
- *Web Service Discovery*: A more complex (and realistic) scenario is to consult WSMX about the set of Web services that satisfy a given goal. This entry point implies a synchronous call – the requester provides a goal and WSMX returns a set of matching Web services.
- *Send Message*: After the decision on which service to use was already made, a conversation involving back and forth messages between the requester and WSMX has to start. The input parameter is a WSML message that contains a set of ontology instances and references to the Web service to be invoked and to the targeted Choreography (if it is available).
- *Store Entity in the Registry*: This entry point provides an interface for storing the WSMO-related entities described in WSML in the repository.

All the incoming and outgoing messages are represented in WSML and consist of either fragments of WSMO ontologies or WSMO entities (Web services, goals, mediators, or ontologies). That is, only WSML is used as WSMX internal data representation and all the necessary adaptations are handled by an adapter framework.

10.2.3.3. WSMX Architecture

To summarize, WSMX architecture (Zaremba *et al.*, 2005) consists of a set of loosely coupled components. Having various decoupled components part of a software system is one of the fundamental principles of a Service Oriented Architecture (SOA). Self-contained components with well-defined functionalities can be easily plugged-in and plugged-out at any time, allowing them to use each others functionalities. Even if WSMX provides a default implementation for all the components in the architecture, following these principles allows a third-party component

---

[8]By execution semantics (or operational semantics) we understand here the formal definition of the operational behavior of the system. It has the role of describing in a formal and unambiguous language how the system behaves.

offering the same functionality (or an enhanced functionality) to be easily plugged-in.

WSMX architecture (see Figure 10.3) provides descriptions of the external interfaces and behaviors for all the components and for the system as a whole. By this, the system's overall functionality is separated from the implementation of particular components. It is worth noting that WSMX accepts as inputs only WSML messages and returns the results as WSML messages as well. In the case of requesters unable to process WSML the Adapter Framework can be used to transform from/ to an arbitrary representation format to/from WSML. For more details about the WSMX infomodel, the reader can check the WSMX code base at Sourceforge[9]. In the future, WSMX intends to support dynamic execution semantics, which means that it will become possible to dynamically load during runtime the intended behavior of the system.

10.3. THE OWL-S APPROACH

OWL-S (2004), part of the DAML program[10], is an OWL-based Web Service Ontology; it aims at providing building blocks for encoding rich semantic service descriptions, in a way that builds naturally upon OWL. Very often is referred to the OWL-S ontology as a *language* for describing services, thus reflecting the fact that it provides a vocabulary that can be used together with the other aspects of the OWL to create service descriptions.

The OWL-S ontology mainly consists of three interrelated sub-ontologies, known as the *profile*, *process model*, and *grounding*. *The profile* is used to express 'what a service does,' for purposes of advertising, constructing service requests, and matchmaking; *the process model* describes 'how it works,' to enable invocation, enactment, composition, monitoring, and recovery; and *the grounding* maps the constructs of the process model onto detailed specifications of message formats, protocols, and so forth (normally expressed in WSDL).

All these sub-ontologies are linked to the top-level concept *Service*, which serves as an organizational point of reference for declaring Web Services; whenever a service is declared, an instance of the *Service* concept is created. As shown in Figure 10.4 below, the properties *presents*, *describedBy*, and *supports* are properties of Service.

The classes ServiceProfile (which identifies the *profile* sub-ontology), ServiceModel (which identifies the *process model* sub-ontology), and ServiceGrounding (which identifies the *grounding* sub-ontology) are the respective ranges of those properties. Each instance of Service will *present*

---

[9]http://sourceforge.net/projects/wsmx/
[10]http://www.daml.org/

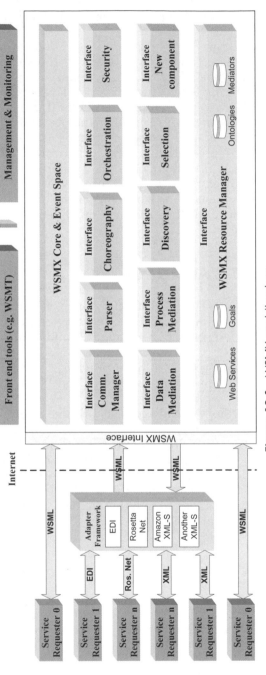

Figure 10.3 WSMX architecture.

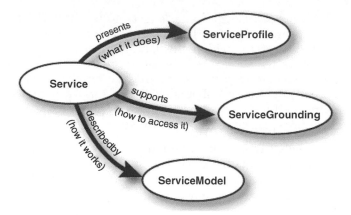

Figure 10.4 Top level elements of OWL-S (OWL-S, 2004).

a ServiceProfile description, be *describedBy* a ServiceModel description, and *support* a ServiceGrounding description.

In the rest of this section we take a closer look at the elements that are part of the OWL-S Service Profile ontology (in Section 10.3.1), and of the OWL-S Service Model ontology (in Section 10.3.2); we do not discuss the OWL-S Grounding ontology in this section as we provide a wider overview of the general problem of SWS Grounding in Section 10.7.

10.3.1. OWL-S Service Profiles

In OWL-S, the Service Profile provides means to describe the services offered by the providers, and the services needed by the requesters. No representation of services is imposed by the Service Profile, but rather, using the OWL sub-classing it is possible to create specialized representations of services that can be used as service profiles. However, for pragmatic reasons, OWL-S provides one possible representation through the class Profile. A service, defined through the OWL-S Profile, is modeled as a function of three basic types of information:

- *The Organization that Provides the Service*: The contact information that refers to the entity that provides the service (e.g., contact information may refer to the maintenance operator that is responsible for running the service, or to a customer representative that may provide additional information about the service, etc.).
- *The Function the Service Computes*: The transformation produced by the service. The functional description includes the inputs required by the service and the outputs generated; the preconditions required by the service and the expected effects that result from the execution of the service.

- *A Host of Features that Specify Characteristics of the Service*: The descriptions of these features include the category of a given service (e.g., the category of the service within the UNSPSC classification system), quality rating of the service (e.g., some services may be very good, reliable, and quick to respond; others may be unreliable, sluggish, or even malevolent), and an unbounded list of service parameters that can contain any type of information (the OWL-S Profile provides a mechanism for representing such parameters).

The most essential type of information presented in the profile, that will play a key role during the discovery of the service, is the specification of what functionality the service provides. The OWL-S Profile emphasizes two aspects of the functionality of the service:

- The Information Transformation: Represented by *inputs* and *outputs* of the service, and
- The State Change produced by the Execution of the Service: Represented by the *preconditions* and *effects* of the service.

No schema to describe inputs/outputs/preconditions/effects (IOPE) instances is provided by the OWL-S Profile. However, such a schema exists in the Process ontology. It is expected that the IOPE's published by the Profile are a subset of those published by the Process, thus it is expected that the Process part of a description will create all the IOPE instances and the Profile instance can simply point to these instances. The properties of the Profile class that the OWL-S Profile ontology defines for pointing to IOPE's are summarized as follows:

- **hasParameter**: Ranges over a Parameter instance of the Process ontology; it's role is solely making domain knowledge explicit.
- **hasInput**: Ranges over instances of Inputs as defined in the Process ontology.
- **hasOutput**: Ranges over instances of type Output, as defined in the Process ontology.
- **hasPrecondition**: Specifies one of the preconditions of the service and ranges over a Precondition instance defined according to the schema in the Process ontology.
- **hasResult**: Specifies one of the results of the service, as defined by the Result class in the Process ontology; it specifies under what conditions the outputs are generated. This parameter also specifies what domain changes are produced during the execution of the service.

10.3.2. OWL-S Service Models

As the OWL-S Profile describes only the overall function the service provides, a detailed perspective on how to interact with the service is needed. This interaction can be viewed as a *process*, and OWL-S defines the ServiceModel subclass in order to provide means to define processes. The view that OWL-S takes on processes is that a process is not necessary a program to be executed, but rather a specification of the ways a client may interact with a service. A process can generate and return some new information based on information it is given and the world state. Information production is described by the inputs and outputs of the process. A process can as well produce a change in the world. This transition is described by the preconditions and effects of the process.

Informally, any process can have any number of *inputs*, representing the information that is, under some conditions, required for starting the process. Processes can have any number of *outputs*, the information that the process provides to the requester. Inputs and outputs are represented as sub-classes of a general class called Parameter; (every parameter has a type, specified using a URI). There can be any number of *preconditions*, which must all hold in order for a process to be successfully started. A process can have any number of *effects*. Outputs and effects can depend on conditions that hold true of the world state at the time the process is performed. Preconditions and effects are represented as logical formulas. OWL-S treats such expressions as literals, either string literals or XML literals. The latter case is used for languages whose standard encoding is in XML, such as SWRL (Horrocks et al., 2003) or RDF (Klyne and Carroll, 2004). The former case is for other languages such as KIF (KIF 1998) and PDDL (PDDL, 1998). Processes are connected to their IOPEs using the following properties:

- **hasParticipant** which ranges over the Participant class.
- **hasInput** which ranges over the Input class.
- **hasOutput** which ranges over the Output class.
- **hasLocal** which ranges over the Local class.
- **hasPrecondition** which ranges over the Condition class.
- **hasResult** which ranges over the Result class.

A process involves at least two parties. One is the client, from whose point of view the process is described, and another is the service that the client deals with. Both the client and the service are referred to as participants; they are directly linked to a process using the *hasParticipant* property. Inputs and outputs specify the data transformation produced by the process; they are directly linked to a process using the *hasInput* and *hasOutput* properties. Inputs specify the information that the process requires for its execution. Inputs may come directly from the client or may come from previous steps of the same process. Outputs specify the information that the process generated after its execution. The presence of a precondition for a process means that the process cannot be performed successfully unless the precondition is true; preconditions are directly linked to a process using the *hasPrecondition* property. The execution of a process may result in changes of the state of the world (effects), and the generation of information by the service (referred to as outputs). Such coupled outputs and effects are not directly linked to a process, but through the term *result* (i.e., through the *hasResult* property).

Although the above properties are common to all processes defined in OWL-S, there can be three types of processes:

- *Atomic* Processes: Description of services that expects one (possibly complex) message and returns one (possibly complex) message in response.
- *Composite* Processes: Processes that maintain some state; each message the client sends advances it through the process.
- *Simple* Processes: processes used as elements of abstraction, that is, a simple process may be used either to provide a view of (a specialized way of using) some atomic process, or a simplified representation of some composite process (for purposes of planning and reasoning).

Atomic processes are similar to the actions a service can perform by engaging it in a single-step interaction; composite processes correspond to actions that require multi-step interactions, and simple processes provide an abstraction mechanism to enable multiple views of the same process. Atomic processes are directly invocable and do not consist of any sub-processes; their execution is a single-step execution (as far as the service requester is concerned), that is they take an input message, do something, and then return their output message. On the other side, composite processes are decomposable into other (atomic, simple, or composite) processes; their decomposition can be specified by using control constructs. The control constructs supported in OWL-S can be summarized as follows:

- *Sequence*: A set of processes to be executed sequentially.
- *Split*: A set of processes to be executed concurrently; it completes as soon as all of its component processes have been scheduled for execution.
- *Split + Join*: Consists of concurrent execution of a bunch of processes with synchronization; it completes when all of its processes have completed.
- *Choice*: Calls for the execution of a single control construct from a given set of control constructs; any of the given control constructs may be chosen for execution.
- *Any-Order*: Allows the process components to be executed in some unspecified order but not concurrently.
- *If-Then-Else*: A control construct that has properties *if-Condition*, *then*, and *else* holding different aspects of the *If-Then-Else*; its intended meaning is as 'Test *If-condition*; if *True* do *Then*, if *False* do *Else*.'
- *Iterate*: A control construct that executes several times a process; this construct makes no assumption about how many iterations are made or when to initiate, terminate, or resume.

- *Repeat-While* and *Repeat-Until*: Both of these iterate until a condition becomes false or true.

A description of a composite process shall not be interpreted as the behavior a service will do, but rather the behavior, or better the set of behaviors the client is allowed to perform by exchanging messages with the service. Furthermore, if the composite process has an overall effect, then the client must perform the entire process in order to achieve that effect.

10.4. THE SWSF APPROACH

Semantic Web Services Framework (SWSF) (SWSF, 2005) is one of the newest approaches for Semantic Web Services, being proposed and promoted by Semantic Web Services Language Committee[11]. (SWSLC) of the Semantic Web Services Initiative[12] (SWSI). It is based on two major components: an ontology and the corresponding conceptual model by which Web services can be described, called *Semantic Web Services Ontology* (*SWSO*) and a language used to specify formal characterisations of Web services concepts and descriptions called *Semantic Web Services Language* (*SWSL*). This section provides a general overview of the two core components of SWSF approach for SWS namely: SWSO – Semantic Web Service Ontology (Section 10.4.1) and SWSL – Semantic Web Service Language (Section 10.4.2).

10.4.1. The Semantic Web Services Ontology (SWSO)

SWSO presents a conceptual model for semantically describing Web services and an axiomatization, formal characterization of this model given in one of the two variants of SWSL: *SWSL-FOL* based on First-Order Logic or *SWSL-Rules* based on Logic programming. The resulting ontologies are called: *FLOWS – First-Order Logic Ontology for Web Services*, which relies on First-Order Logic semantics, and *ROWS – Rule Ontology for Web Services*, which relies on Logic Programming semantics. Since both representations shared the same conceptual model we will focus our overview on FLOWS, the derivation of ROWS from FLOWS being straightforward.

The development of FLOWS ontology was influenced by the OWL-S ontology and the lessons learned from developing this ontology. Another fundamental aspect in the development of FLOWS is the provision of a

---

[11]http://www.daml.org/services/swsl/
[12]http://www.swsi.org/

rich behavioral process model based on Process Specification Language (PSL) (Gruninger, 2003). FLOWS can be seen as an extension/refinement of OWL-S ontology with a special focus on providing interoperability or semantics to existing standards in Web services area (e.g., BPEL, WSDL, etc.) Although there are many similarities between FLOWS and OWL-S ontologies, one important difference is the expressiveness of the under-ling language. FLOWS is based on First-Order logic, which means it has a richer, more expressive, support than OWL-S which is based on OWL-DL.

Being based on First-Order Logic, FLOWS makes use of logic pre-dicates and terms to model the state of the world. Features from situation calculus, like the use of *fluents*, *predicates*, and *terms* which vary over time, were introduced to model the change of the world. Invariant predicates and terms are called in *relations* in SWSO.

The FLOWS ontology consists of three major components: *Service Descriptors*, *Process Model*, and *Grounding*. *The Service Descriptors* are used to provide basic descriptive information about the service. *The Process Model* is used to describe how the service works. The *Grounding* is used to link the semantic, abstract descriptions of the service provided in SWSO to detailed specifications of messages, protocols, and so forth used by Web services.

In the rest of this section we take a closer look at the elements that are part of the FLOWS Service Descriptors (in Section 10.4.1.1), and of the FLOWS Process Model (in Section 10.4.1.2); we do not discuss the FLOWS Grounding in this section as we provide a wider overview of the general problem of SWS Grounding in Section 10.7.

10.4.1.1. Service Descriptors

Service Descriptors are the components of FLOWS ontology which provide basic information about a service. By basic information is meant nonfunctional meta-information and/or provenance information. These kinds of descriptions are often used to support the automation of service related tasks like service discovery. They include information like name, textual description, version, etc, which are properties inherited from the OWL-S Profile. A Service Descriptor may include the following individual properties: (1) *Service Name* – this property refers to the name of the service and may be used as a unique identifier; (2) *Service Author* – this property refers to the authors of the service which can be people or organizations; (3) *Service Contact Information* – this property contains a pointer for the agents or people requiring more information about the service; (4) *Service Contributor* – this property refers to the entity respon-sible for updating the service description; (5) *Service Description* – this property contains the textual description of the service; (6) *Service URL* – this property contains the URL associated with the service; (7) *Service Identifier* – this property contains an unambiguous reference to the service; (8) *Service Version* – this property contains an identifier to the specific version of the service; (9) *Service Release Date* – this property contains the release date of the service; (10) *Service Language* – this property specifies the language of the service; (11) *Service Trust* – this property described the trustworthiness of the service; (12) *Service Subject* – this property refers to the topic of the service; (13) *Service Reliability* – this property contains and entity used to indicate the dependencies of the service; (14) *Service Cost* – this property contains the cost of invocation for the service.

10.4.1.2. Process Model

The Process Model is that part of FLOWS ontology which offers the needed constructs to describe the behavior of the service. The Process Model extends towards the Web services requirements the generic ontology for processes provided by PSL approach, by adding two fundamental elements: (1) the structured notion of atomic process as found in OWL-S and (2) the infrastructure for specifying various forms of data flow. The core part of the PSL extended by FLOWS is called PSL Outer Core and the resulting FLOWS sub-ontology is called FLOWS-Core. The overall extensions to PSL implemented in FLOWS are presented in Figure 10.5.

Based on these extensions FLOWS Process Model ontology can be regarded as a combination of six ontology modules namely:

- *FLOWS-Core*: Introduces the basic notions of activities as activities composed of atomic activities.
- *Control Constraints*: Axiomatize the basic constructs common to work-flow-style process models.

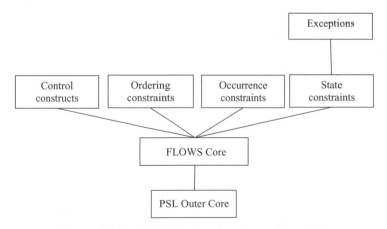

Figure 10.5 FLOWS extensions based on PSL.

- *Ordering Constraints*: Support the specification of activities defined by sequencing properties of atomic processes.
- *Occurrence Constraints*: Support the specification of nondeterministic activities within services.
- *State Constraints*: Support the specification of activities which are triggered by states that satisfies a given condition.
- *Exception Constraints*: Provides support for modeling exceptions.

As part of the FLOWS-Core some basic terms are defined:

- *Service*: A service is defined as an object which has associated a set of service descriptors and an activity that specifies the process model of the service, activities called *service activities*.
- *Atomic Process*: An atomic process is a PSL activity, that is, in general a sub-activity of the activity associated with the service. Associated with each atomic process are (multiple) *input, output, precondition*, and *effects*. The inputs and the outputs are the inputs and outputs of the program which realizes the atomic process. The preconditions are conditions that must be true in the word for the atomic process to be executed. Finally, effects are the side effects of the execution of the atomic process. All these are expressed as First-Order logic formulae.
- *Message*: A message is an object in FLOWS-Core ontology which has associated a *message type* and a *payload* (body).
- *Channel*: A message is an object in FLOWS-Core ontology which holds messages that have been sent and may or may not have received.

10.4.2. The Semantic Web Services Language (SWSL)

SWSL is a language for describing, in a formal way, Web services concepts and descriptions of individual services. SWSL comes in two variants which are based on two well-known formalisms: First-Order Logic and Logic Programming. The two sub-languages are *SWSL-FOL* and *SWSL-Rules*. The design of both languages was driven by compliance with Web principles, like usage of URIs, integration with XML built-in types and XML-compatible namespaces, and import mechanisms. Both languages are *layered languages* where every layer includes a number of new concepts that enhance the modeling power of the language.

SWSL-Rules is a logic programming language which includes features from Courteous logic programs (Grosof, 1999), HiLog (Chen and Kifer, 1993) and F-Logic (Kifer *et al.*, 1995), and can be

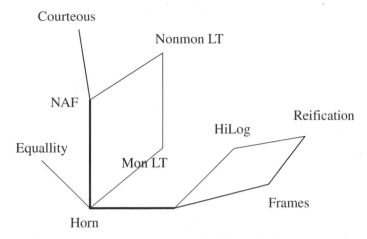

Figure 10.6 The layered structure of swsl-rules (SWSF, 2005).

seen as both specification and implementation language. *SWSL-Rules* language provides support for service-related tasks like discovery, contacting, policy specification, and so on. It is a layered-based languages as shown in Figure 10.6.

The core of the SWSL-Rules language is represented by pure *Horn* sub-set of SWSL-Rules. This subset is extended by adding different features like (1) disjunction in the body and conjunction and implication in the head – this extension is called *monotonic Loyd-Topor (Mon LT)* (Lloyd, 1987), (2) negation in the rule body interpreted as nation as failure – this extension is called *NAF*. Furthermore, the *Mon LT* can be extended by adding quantifiers and implication in the rule body resulting in what is called *nonmonotonic Loyd-Topor (Nonmon LT)* extension. Other envisioned extensions are towards: (1) *Courteous rules (Courteous)* whit two new features: restricted classical negation and prioritized rules, (2) *HiLog* – enables meta-programming, (3) *Frames* – add object oriented features like frame syntax, types, and inheritance, (4) *Reification* – allows rules to be referred and grouped. Finally, *Equality* can be possible extension as well.

SWSL-FOL is a First-Order logic which includes features from HiLog and F-Logic. It has as well a layered structure which is depicted in Figure 10.7

Some of the extensions provided for SWSL-Rules apply for SWSL-FOL as well. The only restriction is that the initial languages should have monotonic semantics. The resulting extensions depicted in Figure 10.7 are *SWSL-FOL + Equality, SWSL-FOL + HiLog,* and *SWSL-FOL + Frame.*

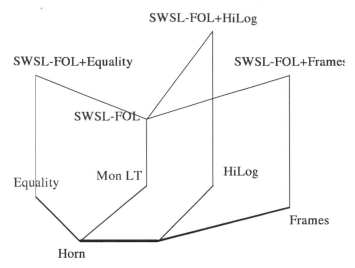

Figure 10.7 The layers of SWSL-FOL and their relationship to SWSL-Rules (SWSF, 2005).

10.5. THE IRS-III APPROACH

IRS-III[13] (Domingue *et al.*, 2004) is a framework and implemented platform which acts as a broker mediating between the goals of a user or client, and available deployed Web services. The IRS uses WSMO as its basic ontology and follows the WSMO design principles. Below we outline additional principles which have influenced the IRS (Section 10.5.1). We then give an overview of the IRS-III architecture (in Section 10.5.2) and present the IRS extensions to WSMO (in Section 10.5.3). In the rest of the section we will use the terms 'IRS' and 'IRS-III' interchangeably.

10.5.1. Principles Underlying IRS-III

IRS-III is based on the following design principles:

- Supporting Capability Based Invocation: IRS-III enables clients (human users or application programs) to invoke a Web service simply by specifying a concrete desired capability. The IRS acts as a broker finding, composing, and invoking appropriate Web services in order to fulfill the request.
- Ease of Use: IRS interfaces were designed so that much of the complexity surrounding the creation of SWS-based applications are hidden. For example, the IRS-III browser hides some of the complexity

---

[13]http://kmi.open.ac.uk/projects/irs/

of underling ontology by bundling up related class definitions into a single tabbed dialog window.

- One Click Publishing: A corollary of the above-design principle. There are many users who have an existing system which they would like to be made available but have no knowledge of the tools and processes involved in turning a stand alone program into a Web service. Therefore, IRS was created so that it supported 'one click' publishing of stand alone code written a standard programming language (currently, we support Java and Lisp) and of applications available through a standard Web browser.

- Agnostic to Service Implementation Platform: This principle is in part a consequent of the one click publishing principle. Within the design of the IRS there is no strong assumptions about the underlying service implementation platform. However, it is accepted the current dominance of the Web services stack of standards and consequently program components which are published through the IRS also appear as standard Web services with a SOAP-based end point.

- Connected to the External Environment: When manipulating Web services, whether manually or automatically, one needs to be able to reason about their status. Often this information needs to be computed on-the-fly in a fashion which integrates the results smoothly with the internal reasoning. To support this we allow functions and relations to be defined which make extra-logical calls to external systems – for example, invoking a Web service. Although, this design principle has a negative effect on ability to make statements about the formal correctness of resulting semantic descriptions, it is necessary because our domain of discourse includes the status of Web services. For example, a user may request to exchange currencies using 'today's best rate.' If our representation environment allows us to encode a current-rate relation which makes an external call to an appropriate Web service or Website then this will not only make life easier for the SWS developer, but also make the resulting descriptions more readable.

- Open: The aim is to make IRS-III as open as possible. The IRS-III clients are based on Java APIs which are publicly accessible. More significantly, components of the IRS-III server are Semantic Web services represented within the IRS-III framework. This feature allows users to replace the main parts of the IRS broker with their own Web services to suit their own particular needs.

- Inspectibility: In many parts of the life cycle of any software system, it is important that the developers are able to understand the design and behavior of the software being constructed. This is also true for SWS applications. This principle is concerned with making the semantic descriptions accessible in a human readable form. The descriptions could be within a plain text editor or within a purpose built browsing or editing environment. The key is that the content and form are easily understandable by SWS application builders.

10.5.2. The IRS-III Architecture

In addition to fulfilling the design principles listed above – especially, supporting capability-based invocation – the IRS-III architecture has been created to link ontology-based descriptions with the components which support SWS activities.

The IRS-III architecture is composed by the main following components: the IRS-III Server, the IRS-III Publisher, and the IRS-III Client, which communicate through a SOAP-based protocol, as shown in Figure 10.8[14].

At the heart of the server is the WSMO library where the WSMO definitions are stored using our representation language OCML (Motta, 1998). The library is structured into knowledge models for WSMO goals, Web services, and mediators. The structure of each knowledge model is similar but typically the applications consist of mediator models importing from relevant goal and Web service models. Following our design principle of inspectibility all information relevant to a Web service is stored explicitly within the library.

Within WSMO a Web service is associated with an interface which contains an orchestration and choreography. Orchestration specifies the control and dataflow of a Web service which invokes other Web services

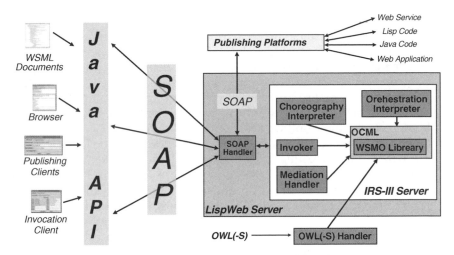

Figure 10.8. The IRS-III server architecture.

---

[14]The IRS-III browser/editor and publishing platforms are currently available at http://kmi.open.ac.uk/projects/irs/

(a composite Web service). Choreography specifies how to communicate with a Web service. The choreography component communicates with an invocation module able to generate the required messages in SOAP format.

A mediation handler provides functionality to interpret WSMO mediator descriptions including running data mediation rules, invoking mediation services, and connecting multiple mediators together. Following from the openness principle above orchestration, choreography, and mediation components are themselves Semantic Web services. At the lowest level the IRS-III Server uses an HTTP server written in lisp (Riva and Ramoni, 1996), which has been extended with a SOAP (XML Protocol Working Group, 2003) handler.

Publishing with IRS-III entails associating a specific web service with a WSMO web service description. When a Web service is published in IRS-III all of the information necessary to call the service, the host, port, and path are stored within the choreography associated with the Web service. Additionally, updates are made to the appropriate publishing platform. The IRS contains publishing platforms to support the publishing of standalone Java and Lisp code, and of Web services. Web applications accessible as HTTP GET requests are handled internally by the IRS-III server.

IRS was designed for ease of use, in fact a key feature of IRS-III is that Web service invocation is capability driven. The IRS-III Client supports this by providing a goal-centric invocation mechanism. An IRS user simply asks for a goal to be solved and the IRS broker locates an appropriate Web service semantic description and then invokes the underlying deployed Web service.

10.5.3. Extension to WSMO

The IRS-III ontology is currently based on the WSMO conceptual model with a number differences mainly derived from the fact that in IRS-III the aim is to support capability driven Web service invocation. To achieve these goals, Web services are required to have input and output roles. In addition to the semantic type the soap binding for input and output roles is also stored. Consequently, a goal in IRS-III has the following extra slots *has-input-role, has-output-role, has-input-role-soap-binding, and has-output-role-soap-binding*.

Goals are linked to Web services via mediators. More specifically, the WG Mediators found in the used-mediator slot of a Web service's capability. If a mediator associated with a capability has a goal as a source, then the associated Web service is considered to be linked to the goal.

Web services which are linked to goals 'inherit' the goal's input and output roles. This means that input role definitions within a Web service are used to either add extra input roles or to change an input role type.

When a goal is invoked the IRS broker creates a set of possible contender Web services using the WG Mediators. A specific web service is then selected using an applicability function within the assumption slot of the Web service's associated capability. As mentioned earlier the WG Mediators are used to transform between the goal and Web service input and output types during invocation.

In WSMO the mediation service slot of a mediator may point to a goal that declaratively describes the mapping. Goals in a mediation service context play a slightly different role in IRS-III. Rather than describing a mapping goals are considered to have associated Web services and are therefore simply invoked.

IRS clients are assumed to be able to formulate their request as a goal instance. This means that it is only required choreographies between the IRS and the deployed Web services. In IRS-III choreography execution thus occurs from a client perspective (Domingue *et al.*, 2005), that is to say, to carry out a Web service invocation, the IRS executes a web service *client choreography* which sends the appropriate messages to the deployed Web service. In contrast, currently, WSMO choreography describes all of the possible interactions that a Web service can have.

10.6. THE WSDL-S APPROACH

WSDL-S (Akkiraju *et al.*, 2005) proposes a mechanism to augment the Web service functional descriptions, as represented by WSDL (WSDL, 2005), with semantics. This work is a refinement of an initial proposal developed by the Meteor-S group, at the LSDIS Lab[15], Athens, Georgia.

In this section we briefly present the principles WSDL-S is based on (in Section 10.6.1), and we shortly describe the extensibility elements used and the annotations that can be created (in Section 10.6.2).

10.6.1. Aims and Principles

Starting from the assumption that a semantic model of the Web service already exists, WSDL-S describes a mechanism to link this semantic model with the syntactical functional description captured by WSDL. Using the extensibility elements of WSDL, a set of annotations can be created to semantically describe the inputs, outputs, and the operation of

---

[15]See http://lsdis.cs.uga.edu/.

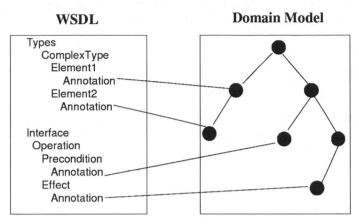

Figure 10.9 Associating semantics to WSDL elements (Akkiraju *et al.*, 2005).

a Web service. By this the semantic model is kept outside WSDL, making the approach agnostic to any ontology representation language (see Figure 10.9).

The advantage of such an approach is that it is an incremental approach, building on top of an already existing standard and taking advantage the already existing expertise and tool support. In addition, the user can develop in WSDL in a compatible manner both the semantic and operational level aspects of Web services.

WSDL-S work is guided by a set of principles, the most important of them being listed below:

- *Build on Existing Web Services' standards*: Standards represent a key point in creating integration solutions and as a consequence, WSDL-S promotes an upwardly compatible mechanism for adding semantics to Web services.
- *Annotations Should be Agnostic to the Semantics Representation Language*: Different Web service providers could use different ways of representing the semantic descriptions of their services and furthermore, the same Web service provider can choose more than one representation form in order to enable its discovery by multiple engines. Consequently, WSDL-S does not prescribe what semantic representation language should be used and allows the association of multiple annotations written in different semantic representation languages.
- *Support Annotation of XML Schema Data Type*: As XML Schema is an important data definition format and it is desirable to reuse the existing interfaces described in XML, WSDL-S supports the annotation of XML Schemas. These annotations are used for adding

semantics to the inputs and outputs of the annotated Web service. In addition, an important aspect to be considered is the creation of mappings between the XML Schema complex types and the corresponding ontological concepts. As WSDL-S does not prescribe an ontology language, the mapping techniques would be directly dependent of the semantic representation language chosen.

In the next subsection we present in more details the extensibility elements of WSDL and how they can be used in annotating the inputs, outputs, and operations of Web services.

10.6.2. Semantic Annotations

WSDL-S proposes five extensibility elements to be used in annotating the inputs, outputs, and operations of Web services:

- **modelReference:** Extension element that denotes a one-to-one mapping between schema elements and concepts from the ontology;
- **schemaMapping:** Extension attribute that can be added to XSD elements or complex types to associate them with an ontology (used for one-to-many and many-to-one mappings);
- **precondition:** Extension element (child of the *operation* element) used to point to a combination of complex expressions and conditions in the ontology, that have to hold before the execution of the Web service's operation;
- **effects:** Similar with preconditions, with the difference that the conditions in the ontology have to hold after the execution of the Web service's operation.
- **category:** Extension attribute of the *interface* element that points to categorization information that can be used for example when publishing the Web service.

Each of these elements can be used to create annotations; in the rest of this section we briefly describe each type of annotations, pointing to the extensibility elements used.

10.6.2.1. Annotating the Input and Output Elements

If the input or the output is a *simple type* it can be annotated using the extensibility of the XML Schema element: the *modelReference* attribute is used to associate annotations to the element.

If the input or the output is a *complex type* two strategies can be adopted: bottom level annotation and top level annotation. In *bottom level annotation* all the leaf elements can be annotated with concepts from the ontology. The modelReference attribute is used here in a similar manner as above. While this method is simple, it makes the assumption that there is one-to-one correspondence between the elements from the XML Schema and the concepts from the ontology. In the case of one-to-many or many-to-one correspondences *top level annotation* method has to be used. Although it is a more complex method, its advantage is that it allows for complex mappings to be specified between the XML element and the domain ontology. The semantic of the elements in the complex type is captured by using the *schemaMapping* attribute.

10.6.2.2. Annotating the Operation Elements

The operations of a Web service can be annotated with *preconditions*, which represent a set of assertions that must hold before the execution of that operation. The precondition extension element is defined as follows:

- **/precondition:** Denote the precondition for the parent operation;
- **/precondition/@name:** Uniquely identifies the precondition among the other preconditions defined in the WSML document;
- **/precondition/@modelReference:** Points to that parts of the semantic model that describes the precondition of this operation;
- **/precondition/@expression:** Contains the precondition associated to the parent operation. Its format directly depends on the semantic representation language used. The two ways of specifying the precondition assertions, */precondition/*@expression, and */precondition/*@modelReference are mutually exclusive.

For each operation there is only one precondition allowed. This restriction is adopted as an attempt to keep the specification simple. If one needs more than one precondition, the solution is to define in the domain ontology the complex expressions and conditions and to point to them using the modelReference attribute.

The *effects* define the result of invoking a particular operation. The *effect* element is defined in a similar manner as the precondition (see above), and it is allowed to have one or more effects associated with one operation.

10.6.2.3. Service Categorization

Adding categorization information to the Web service can be helpful in the discovery process. That is, by categorizing the published Web services can narrow the range of the candidate Web services. Multiple category elements can be used to state that a Web service falls in multiple categories as one category elements specifies one categorization.

10.7. SEMANTIC WEB SERVICES GROUNDING: THE LINK BETWEEN THE SWS AND EXISTING WEB SERVICES STANDARDS

As we have pointed in the previous sections, the ultimate aim of SWS – automatic execution of tasks like discovery, negotiation, composition, invocation of services – requires semantic description of various aspects of Web services. For example, the process of Web service discovery can be automated if we have a machine-processable description of what the user wants (a user goal) and what the available services can do (service capabilities). We call this kind of information *semantic description* of Web services.

However, currently deployed Web services are generally described only on the level of syntax, specifying the structure of the messages that a service can accept or produce. In particular, Web Service Description Language (WSDL, 2005) describes a Web service interface as a set of operations where an operation is only a sequence of messages whose contents are constrained with XML Schema (2004). We call this the *syntactic description* of Web services.

Certain tasks require that semantic processors have access to the information in the syntactic descriptions, for example to invoke a discovered service, the client processor needs to know how to serialize the request message. Linking between the semantic and the syntactic description levels is commonly called *grounding*. In order for SWS to be widely adopted, they must provide mechanisms that build on top of existing, widely adopted technologies. In this Section we look at such mechanisms and discuss the general issues of Semantic Web Service grounding (in Section 10.7.1); we also identify two major types of grounding, so in Section 10.7.2 we talk about data grounding and in Section 10.7.3 we talk about grounding behavior descriptions.

10.7.1. General Grounding Uses and Issues

As we have shown in the previous sections, most of the existing approaches to Semantic Web Services describe services in terms of their functional and behavioral properties, using logics-based (ontological) formalism. First, to enable Web service discovery and composition, SWS frameworks need to describe what Web services do, that is, service capabilities. Second, to make it possible for clients to determine how to communicate with discovered services, the interfaces of the services need to be described. The description of a service interface must be sufficient for a client to know how to communicate successfully with the Web service; in particular a service interface must describe the messages and the networking details. For interoperability with existing Web services and infrastructures, interface description is based on WSDL. The glue between the semantic interface description and WSDL is called grounding. WSDL models a service interface as a set of operations representing message exchanges. Message contents are specified abstractly as XML Schema element declarations, and then WSDL provides so-called binding information with the specific serialization and networking details necessary for the messages to be transmitted between the client and the service.

On the data level, Semantic Web Service frameworks model Web services as entities that exchange semantic (ontological) data. The grounding must provide means to represent that semantic data as XML messages to be sent over the network (according to serialization details from the WSDL binding), and it must also specify how received XML messages are interpreted as semantic data. We investigate this aspect of grounding below in Section 10.7.2.

A Web service interface in WSDL contains a number of operations. Within an operation, message ordering is prescribed by the message exchange pattern followed by the operation. WSDL does not specify any ordering or dependencies between operations, so to automate Web service invocation, a Semantic Web Service interface must tell the client what particular operations it can invoke at a specific point in the interaction. We call this the choreography model, and very different choreography models are used by the known SWS frameworks. However, since they all ground to WSDL, the grounding must tie the choreography model with WSDL's simple model of separate operations. This aspect of grounding is further detailed in Section 10.7.3.

We now know what kind of information must be specified in grounding, and we have to choose where to place that information, assuming that the semantic description is in a document separate from the WSDL. There are three options for placing grounding information:

- putting grounding in the semantic description,
- embedding grounding within WSDL,
- externalizing grounding in a third document.

Putting grounding within the semantic description format makes it straightforward to access the grounding information from the semantic data, which follows the chronological order of SWS framework tasks – discovery only uses the semantic data, and then invoking the discovered service needs grounding. For example, this approach is currently taken by both WSMO and OWL-S.

On the other hand, putting grounding information directly in WSDL documents (option 2) can enable discovering semantic descriptions in WSDL repositories, for example, in UDDI (UDDI, 2004). This approach is taken by WSDL-S (Akkiraju *et al.*, 2005), a specification of a set of WSDL hooks that can be used with any SWS modeling framework. WSDL-S itself is not, however, a full SWS framework. An externalized grounding (outside both WSDL and the semantic descriptions) does not provide either side (semantic or syntactic) with easy access to the grounding information, but it may introduce even more flexibility for reuse.

However, externalized grounding is not supported by any current specifications. We must note that the options listed above are not exclusive, so grounding information can be put redundantly both in the semantic description document and in the WSDL, for example, so that it is available from both sides. This could be done, for example, by using the native grounding mechanism in WSMO to point to WSDL and at the same time annotating the WSDL with WSDL-S elements pointing back to WSMO.

10.7.2. Data Grounding

Web services generally communicate with their clients using XML messages described with XML Schema. On the semantic level, however, Web service inputs and outputs are described using ontologies. A semantic client then needs grounding information that describes how the semantic data should be written in an XML form that can be sent to the service, and how XML data coming back from the service can be interpreted semantically by the client. In other words, the outgoing data must be transformed from an ontological form to XML and, conversely, the incoming data must be transformed from XML to an ontological form.

Since the semantics of XML data is only implicit, at best described in plain text in the specification of an XML language, a human designer may be required to specify these data transformations so that they can be executed when a semantic client needs to communicate with a syntactic Web service.

In Figure 10.10 we propose a way of distinguishing between data grounding approaches. The figure shows a fragment of the ontology of the semantic description of an example Web service in the upper right corner and the XML data described in WSDL in the lower left corner. The three different paths between the XML data quadrant and the semantic quadrant present different options where the transformations can be implemented: on the XML level, on the semantic level, and a direct option spanning the two levels (Note that we use WSMO terms in the figure but it applies equally to OWL ontologies).

First, since Semantic Web ontologies can be serialized in XML, an XSLT (XSLT, 1999) (or similar) transformation can be created between the XML data and the XML serialization of the ontological data. This approach is very simple and it uses proven existing technologies, but it has a notable

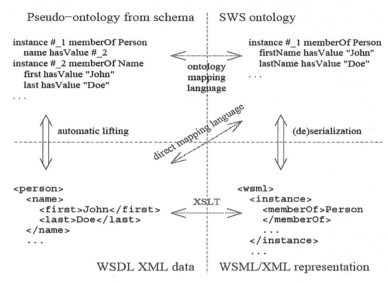

Pseudo–ontology from schema | SWS ontology

instance #_1 memberOf Person
   name hasValue #_2
instance #_2 memberOf Name
   first hasValue "John"
   last hasValue "Doe"
. . .

ontology
mapping
language

instance #_1 memberOf Person
   firstName hasValue "John"
   lastName hasValue "Doe"
. . .

automatic lifting

(de)serialization

direct mapping language

```
<person>
   <name>
      <first>John</first>
      <last>Doe</last>
   </name>
   . . .
```

XSLT

```
<wsml>
   <instance>
      <memberOf>Person
      </memberOf>
      . . .
   </instance>
   . . .
```

WSDL XML data | WSML/XML representation

Figure 10.10   Data grounding approaches.

disadvantage: an XML representation of ontological data (like RDF/XML or WSML/XML) is often an unpredictable mixture of hierarchy and interlinking, as ontological data is not structured according to XML conventions (we say ontological data is not *native* XML data), so creating robust XSLT transformations for both directions may be a considerable task when working with complex data structures. With simple data, however, this problem is negligible, and since XSLT processors are readily available and many XML-savvy engineers have some XSLT experience, this approach is an ideal initial candidate for data grounding. In case the XML serialization of the ontological data is also suitable for the WSDL of a particular Web service, the transformation can be avoided. This approach does not require any human designer to create grounding transformations, which may be a significant saving of effort. On the other hand, XML serializations of ontological data are not native XML data, so they may be hard to comprehend or hard to process by XML tools, and services that use this grounding approach may not integrate well with nonSemantic Web services.

Second, an *ad hoc* ontology can be generated from the XML Schema present in the WSDL, with automatic *lifting/ lowering* between the XML data and their equivalent in the *ad hoc* ontology. Then a transformation using an ontology mapping language can be designed to get to the target ontology. In case the semantic description designer finds this generated ontology suitable for describing this particular Web service, this grounding can be fully automatic. On the other hand, if the generated *ad hoc* ontology is not sufficient or suitable, grounding involves an additional transformation between instances of the *ad hoc* ontology and instances of the target ontology used by the service description. This transformation can be implemented using ontology mediation approaches. Similarly to the XSLT approach, the *ad hoc* ontology approach has the benefit of reusing existing transformation technologies (ontology mediation in this case), and it also has the disadvantage that the generated *ad hoc* ontology is not a native ontology (it is structured as a restrictive schema for data validation, as opposed to a descriptive ontology for knowledge representation), and this ontology can lack or even misrepresent semantics that are only implied in the XML. This can complicate the task of mediating between the *ad hoc* ontology and the target ontology describing the service in a similar way as the nonnative XML data can complicate the XSLT transformation.

Finally, a direct approach for mapping between XML data and the target semantic data can be envisioned. Although we are not aware of any work in this direction in either of the SWS frameworks, we envision a third option that transforms between the XML data and the ontological data directly, using a specific transformation language. While a new transformation language would have to be devised for this approach, it could be optimized for the common transformation patterns between native ontological data and native XML, and so the manually created mappings could be simpler to understand, create, and manage, than in the previous approaches. Therefore, this approach should be considered if the disadvantages of the others prove substantial.

10.7.3. Behavioural Grounding

For the purpose of our discussion on the behavioral grounding, we define a *choreography model* of a Semantic Web Service framework as such part of the semantic description, that allows the client semantic processor to know what messages it can send or receive at any specific point during the interaction with a service. Choreography descriptions have other uses as well, for example, detecting potential deadlocks, but these uses are out of scope of this discussion.

Because Semantic Web Services reuse WSDL, their choreography models must be tied to its simple model of separate operations, each one representing a limited message exchange. The ordering of messages within any single operation is defined by the operation's message exchange pattern, so the choreography model must specify in what sequence the operations can be invoked.

Choreography can be described *explicitly*, using some language that directly specifies the allowed order of operations. Conversely, choreography can also be described *implicitly* (or *indirectly*), by specifying the conditions under which operations can be invoked, and the effects of such invocations, and sometimes also the inputs and outputs, which are conditions on the data that comes in and out of the service. Inputs, outputs, preconditions, and effects are together commonly known as IOPEs. For example, OWL-S and WSDL-S both allow IOPEs to be specified on the level of WSDL operations. WSMO specifies the conditions and effects using abstract state machines.

In the case of *implicit choreography*, IOPEs are usually specified on the level of WSDL operations. With this information, known AI planning techniques (Nau *et al.*, 2004) can be used to find a suitable ordering of the operations, based on the initial conditions and the end goal. In other words, the semantic client processor gets the description of IOPEs for all the available operations and then it plans the actual sequence it will use. The main benefit of the *implicit choreography* approach is its significant flexibility and dynamism, as different operation sequences can be dynamically chosen depending on the goal of a particular client, and the sequence can even be replanned in the middle of a run if some conditions unexpectedly change. However, planning algorithms usually have high computational complexity and require substantial resources, especially if there is a large number of available operations. In situations where the cost of AI planning is a problem, explicit choreographies can be precomputed (or designed) for the supported goals, and these choreographies can then be described explicitly.

On the other side, an *explicit choreography* description specifies, using some kind of process modeling language, the sequences of operations that are allowed on a particular Web service. The client processor must be able to discover the choreography description and then it simply follows what is prescribed. For example, OWL-S can describe choreographies explicitly with so-called composite processes, that is, compositions of atomic processes. The composition ontology is based on various works in workflow and process modeling, and it contains constructs for describing the well-known composition patterns like sequence, conditional execution, and iteration. A client processor following an OWL-S composite process will simply execute the composition constructs, and grounding information will only be needed on the level of atomic processes. No other grounding information is necessary. Alternatively to a SWS-specific composition language, a Web service choreography can be described with industrial languages (i.e., languages developed by the companies heavily involved in Web services standardization) like WSCI (WSCI, 2002)/WS-CDL (WS-CDL, 2004). In this case it would be the goal of grounding simply to point from a Semantic Web Service description to the appropriate choreography document in WS-CDL or any other suitable language. WSMO does not currently support any explicit choreography description, but we expect that if the need arises, an industrial choreography language can easily be adopted, as the grounding requirement of this approach is minimal – the pointer to a WS-CDL document, for example, can be implemented as a nonfunctional property of a service description in WSMO.

10.8. CONCLUSIONS AND OUTLOOK

Semantic Web Services constitute one of the most promising research directions to improve the integration of applications within and across enterprise boundaries. In this context, we provided in this chapter an overview of the most important approaches to SWS and pointed out the main concepts that they define. Although a detailed comparison of all the approaches is out of scope of this chapter, we argue that, in order for SWS to succeed, a fully fledged framework needs to be provided: starting with a conceptual model, continuing with a formal language to provides formal syntax and semantics (based on different logics in order to provide different levels of logical expressiveness) for the conceptual model, and ending with an execution environment that glue all the components that use the language for performing various tasks that would eventually enable automation of service.

Amongst the presented approaches, only the WSMO Approach tackles, in a unifying manner, all the aspects of such a framework, and potentially provides the conceptual basis and the technical means to realize Semantic Web Services: it defines a *conceptual model* (WSMO) for defining the basic concepts of Semantic Web Services, a *formal language* (WSML) which provides a formal syntax and semantics for WSMO by offering different variants based on different logics in order to provide different levels of logical expressiveness (thus allowing different trade offs between expressivity and computability), and an *execution environment* (WSMX) which provides a reference implementation for WSMO and interoperation of Semantic Web Services.

The OWL-S Approach is based on OWL; OWL was not developed with the design rationale in mind to define the semantics of processes that require rich definitions of their functionality, thus inherently limiting the expressivity of OWL-S. WSMO/WSML tries to overcome this limitation by providing different layers of expressivity, thus allowing rich definitions of Web services. Moreover, OWL-S inherits some of the drawbacks of OWL (de Brujin, 2005a): lack of proper layering between RDFS and the less expressive species of OWL, and the lack of proper layering between OWL DL and OWL Lite on the one side and OWL Full on the other. OWL-S provides the choice between several other languages, for example, SWRL, KIF, etc. By leaving the choice of the language to be used to the user, OWL-S contributes to the interoperability problem rather than solving it. In OWL-S, the interaction between the inputs and outputs, which have been specified as OWL classes and the logical expressions in the respective languages, is not clear. OWL-S does not make any explicit distinction between Web service communication and cooperation. WSMO makes this distinctions in terms of Web service choreography and orchestration, thus apply the principle of separation of concerns between communication and cooperation, and making the conceptual modeling more clear. OWL-S does not explicitly consider the heterogeneity problem in the language itself, treating it as an architectural issue, that is, mediators are not an element of the ontology but repart of the underlying Web service infrastructure. WSML provides an integrated language framework for the description of both the ontologies and the services. Furthermore, the logical language used for the specification of Web Service preconditions and postconditions is an integral part of the language, thus the overall web service description and the logical expressions which specify the pre- and postconditions are connected for free.

The SWSF Approach can be seen as an attempt to extend on the work of OWL-S, to incorporate a variety of capabilities not within the OWL-S goals. A difference between FLOWS – the ontology part of SWSF and OWL-S is the expressive power of the underlying language. FLOWS is based On first-Order logic, which means that it can express considerably more than can be expressed using, for example, OWL-DL.

The use of First-Order logic enables a more refined approach than possible in OWL-S to representing different forms of data flow that can arise in Web services. Another difference is that FLOWS tries to explicitly model more aspects of Web services than OWL-S; this includes the fact that FLOWS can readily model process models using a variety of different paradigms and data flow between services, which is achieved either through message passing or access to shared *fluents*. Although the SWSF Approach seems to tackle both conceptual modeling, as well as language issues, it is very unclear how all the paradigms part of this approach work together. Moreover, the purpose of FLOWS was to develop of First-Order logic ontology for Web services, and not a Web language, for example, FLOWS does not even use URIs to specify their concepts.

Amongst all the approaches presented in this chapter, only the IRS-III Approach is integrated with the WSMO Approach in the sense that IRS-III uses WSMO as its underlying epistemological framework. Within IRS-III the stress is on creating a capability-based *broker* (facilitating the invocation of Web services through WSMO goals), ease-of-publication being able to turn standalone code into a SWS through a single simple dialog, and tightly coupling the semantic descriptions with deployed Web services (e.g., semantic concepts and relations can be implemented as Web services). Ongoing work continues to align the two approaches.

The WSDL-S Approach follows a much more technology-centerd approach, not providing a conceptual model for the description of Web services and their related aspects, but rather being a bottom up approach (annotating existing standards with metadata) than a top down, complete, solution to the integration problem. WSDL-S can actually be used to represent a grounding mechanism for WSMO. Being ontology language agnostic, WSDL-S allows Web service providers to directly annotate their services using WSML. That is, *modelReference* attributes can point to concepts from WSML ontologies and the expressions in precondition or effects can be directly described in WSML.

Finally, we highlighted the importance and described possible approaches to *grounding* in the context of Semantic Web Services, as a key enabler for the adoption of SWS technologies to a wide audience.

With the W3C[16] submissions of WSMO (http://www.w3.org/Submission/WSMO), OWL-S (http://www.w3.org/Submission/OWL-S), and SWSF (http://www.w3.org/Submission/SWSF), it is expected that all these approaches to converge in the future in the form of a W3C activity in the area of Web services, in order to provide a standardized framework for Semantic Web Services.

REFERENCES

Akkiraju R, Farrell J, Miller J, Nagarajan M, Schmidt M, Sheth A, Verma, K. 2005.Web Service Semantics – WSDL-S. Technical note, April 2005. Available at http://lsdis.cs.uga.edu/library/download/WSDL-S-V1.html.

Alonso G, Casati F, Kuno H, Machiraju V. 2004. *Web Services: Concepts, Architecture and Applications*. Springer-Verlag.

Arkin A, Askary S, Fordin S, Jekeli W, Kawaguchi K, Orchard D, Pogliani S, Riemer K, Struble S, Takacsi-Nagy P, Trickovic I, Zimek S. 2002. Web Services Choreography Interface 1.0, June 2002. Available at http://www.w3.org/ TR/wsci/.

Baida Z, Gordijn J, Omelayenko B, Akkermans H. 2005. A shared service terminology for online service provisioning. In *ICEC '04: Proceedings of the 6th international conference on Electronic commerce.*

Berners-Lee T, Hendler J, Lassila O. 2001. The semantic web. *Scientific American* 284(5):34–43.

Chen W, Kifer M, Warren DS. 2003. *HiLog: A foundation for higher-order Logic Programming. Journal of Logic Programming* 15:3, 187–230.

Cimpian E, Moran M, Oren E, Vitvar T, Zaremba M. 2005. Overview and Scope of WSMX. Technical report, WSMX Working Draft, http://www.wsmo.org/TR/d13/d13.0/v0.2/, February 2005.

de Bruijn J, editor. 2005. The Web Service Modeling Language WSML. WSMO Deliverable D16, WSMO Final Draft v0.2, 2005, http://www.wsmo.org/TR/d16/d16.1/v0.2/.

de Bruijn J, Polleres A, Lara R, Fensel D. 2005a Owl DL vs. OWL FLight: Conceptual modeling and reasoning for the semantic web. In *Proceedings of the 14th International World Wide Web Conference.*

Dean M, Schreiber G, editors. 2004. OWL Web Ontology Language Reference. W3C Recommendation 10 February 2004.

Domingue J, Cabral L, Hakimpour F, Sell D, Motta E. 2004. Irs-iii: A platform and infrastructure for creating WSMO-based semantic web services. In *Proceedings of the Workshop on WSMO Implementations* (*WIW 2004*), Frankfurt, Germany, September 2004. CEUR.

Domingue J, Galizia S, Cabral L. 2005. Choreography in IRS-III- Coping with Heterogeneous Interaction Patterns in Web Services. In Proceedings of 4th *International Semantic Web Conference* (ISWC 2005), 6–10 November, Galway, Ireland.

---

[16] http://www.w3.org/

Fensel D, Bussler C. 2002. The Web Service Modeling Framework WSMF. *Electronic Commerce Research and Applications* 1(2): XX.

Grosof BN. July 1999. *A Courteous Compiler From Generalized Courteous Logic Programs To Ordinary Logic Programs*. IBM Report included as part of documentation in the IBM CommonRules 1.0 software toolkit and documentation, released on http://alphaworks.ibm.com. Also available at: http://ebusiness.mit.edu/bgrosof/#gclp-rr-99k.

Grosof BN, Horrocks I, Volz R, Decker S. 2003. Description logic programs: Combining logic programs with description logic. In *Proceedings of International Conference on the World Wide Web (WWW-2003)*, Budapest, Hungary.

Gruninger M. 2003. A guide to the ontology of the process specification language. In *Handbook on Ontologies in Information Systems*, Studer R, Staab S (eds). Springer-Verlag.

Hakimpour F, Domingue J, Motta E, Cabral L, Lei Y. 2004. Integration of OWL-S into IRS-III. In *Proceedings of the first AKT Workshop* on *Semantic Web Services*.

Hoare CAR. 1969. An axiomatic basis for computer programming. *Communications of the ACM* 12(10):576–580.

Horrocks I, Patel-Schneider PF, Boley H, Tabet S, Grosof B, Dean M. 2003. Swrl: A semantic web rule language combining owl and ruleml. Available at http://www.daml.org/2003/11/swrl/.

Keller U, Lara R, Polleres A, editors. 2004. WSMO Web Service Discovery. WSMO Deliverable D5.1, WSMO Working Draft, latest version available at http://www.wsmo.org/2004/d5/d5.1/.

Knowledge Interchange Format (KIF). 1998. Draft proposed American National Standard (dpans). Technical Report 2/98-004, ANS. Also at http://logic.stanford.edu/kif/dpans.html.

Kifer M, Lara R, Polleres A, Zhao C, Keller U, Lausen H, Fensel D. 2004. A logical framework for web service discovery. In ISWC 2004 Workshop on Semantic Web Services: Preparing to Meet the World of Business Applications, Hiroshima, Japan, November 2004.

Kifer M, Lausen G, Wu J. 1995. Logical foundations of object-oriented and frame-based languages. *JACM* 42(4):741–843.

Klyne G, Carroll JJ. 2004. Resource description framework (rdf): Concepts and abstract syntax. W3C Recommendation. Available at http://www.w3.org/TR/2004/REC-rdf-concepts-20040210.

Lloyd JW. 1987. *Foundations of Logic Programming (second, extended edition)*. Springer series in symbolic computation. Springer-Verlag: New York.

McIlraith S, Son TC, Zeng H. 2001. Semantic Web Services. In IEEE Intelligent Systems. *Special Issue on the Semantic Web* 16(2):46–53.

Motta E. 1998. An Overview of the OCML Modelling Language. In *proceedings of the 8th Workshop on Knowledge Engineering Methods and Languages* (KEML '98).

Nau D, Ghallab M, Traverso P. 2004. *Automated Planning: Theory & Practice*. Morgan Kaufmann Publishers Inc.: San Francisco, CA, USA.

The OWL Services Coalition. OWL-S 1.1. 2004. Available from http://www.daml.org/services/owl-s/1.1/, November 2004.

Object Management Group Inc. (OMG). 2002. Meta object facility (MOF) specification v1.4.

PDDL–The Planning Domain Definition Language V. 2. 1998. Technical Report, report CVC TR-98-003/DCS TR-1165, Yale Center for Computational Vision and Control.

Preist C. 2004. A conceptual architecture for semantic web services. In *3rd International Semantic Web Conference (ISWC2004)*. Springer Verlag: XX, November 2004.

Riva A, Ramoni M. 1996. LispWeb: A Specialised HTTP Server for Distributed AI Applications. *Computer Networks and ISDN Systems* 28: 7–11, 953–961.

Roman D, Lausen H, Keller U, editors. 2005. Web Service Modeling Ontology (WSMO). WSMO Working Draft D2v1.2, April 2005. Available from http://www.wsmo.org/TR/d2/v1.2/.

Semantic Web Services Framework. SWSF Version 1.0. 2005. Available from http://www.daml.org/services/swsf/1.0/.

UDDI Version 3.0.2. 2004. OASIS Standard, October 2004. Available at http://www.oasis-pen.org/committees/uddi-spec/doc/spec/v3/uddi-v3.0.2-20041019.htm.

Weibel S, Kunze J, Lagoze C, Wolf M. 1998. Dublin core metadata for resource discovery. RFC 2413, IETF, September 1998.

Web Services Choreography Description Language (WS-CDL) Version 1.0. 2004. Working Draft, W3C, December 2004. Available at http://www.w3.org/TR/2004/WD-ws-cdl-10-20041217/.

Web Services Description Language (WSDL) Version 2.0. 2005. Last Call Working Draft, W3C WS Description Working Group, August 2005. Available at http://www.w3.org/TR/2005/WD-wsdl20-20050803.

XML Protocol Working Group. 2003. Soap version 1.2. Technical report, June 2003. W3C Recommendation. Available from http://www.w3.org/TR/2003/REC-soap12-part0-20030624/

XML Schema Part 1. 2004. Structures Recommendation, W3C, October 2004. Available at http://www.w3.org/TR/xmlschema-1/.

XSL Transformations. 1999. Recommendation, W3C, November 1999. Available at http://www.w3.org/TR/xslt.

Zaremba M, Oren E. 2005. WSMX Execution Semantics. Technical report, WSMX Working Draft, http://www.wsmo.org/TR/d13/d13.2/v0.2/, April 2005.

Zaremba M, Moran M, Haselwanter T. 2005. WSMX Architecture. Technical report, WSMX Working Draft, http://www.wsmo.org/TR/d13/d13.4/v0.2/, April 2005.

11

Applying Semantic Technology to a Digital Library

Paul Warren, Ian Thurlow and David Alsmeyer

11.1. INTRODUCTION

The extensive deployment of digital libraries over the last two decades is hardly surprising. They offer remote access to articles, journals and books with many users able to access the same document at the same time. Through the use of search engines, they make it possible to locate specific information more rapidly than ever is possible in physical libraries. Scholars, and others, are able to access rare and precious documents with no danger of damage. However, challenges remain if the full benefits are to be realised. Interoperability between different libraries, or even between different collections in the same library, is a problem. At the semantic level, different schemas are used by different library databases. Search and retrieval needs to be made easier, in part by offering each user a unified view of the naming of digital objects across libraries. User interfaces need to be improved, in particular to face the challenge of large information collections. This chapter describes the state-of-the art in digital library research, and in particular the application of semantic technology to confront the challenges posed.

Subsequent sections go into more detail, but it is clear that the challenges described above align closely with the goals of semantic knowledge technology. The ontology mediation techniques described in Chapter 6 are specifically motivated by the challenge of interoperability between heterogeneous data sets, and of providing a common

Semantic Web Technologies: Trends and Research in Ontology-based Systems
John Davies, Rudi Studer, Paul Warren   © 2006 John Wiley & Sons, Ltd

view to those data sets. As discussed in Chapter 8, semantic information access offers improved ways to search for and browse information and, through an understanding of the relationship between documents, to improve the user interface. Semantic access to information depends in turn on the supporting technologies described in the preceding chapters; while the creation and maintenance of ontologies in digital libraries create problems of ontology management which require new insights into ontology engineering.

The discussion is illustrated with a particular case study in which semantic knowledge technology is being introduced into the BT digital library. This provides an opportunity not just to trial the feasibility of the technology, but also to gauge the users' reactions and better understand their requirements. Finally, it should be remembered that digital libraries are themselves a particular form of content management application. Much of what is being learned here is relevant in the wider context of intelligent content management. To underline this point, the chapter concludes by looking beyond the current concept of the digital library to how semantic technology will change the way in which information is published, thereby changing the whole concept of a library.

11.2. DIGITAL LIBRARIES: THE STATE-OF-THE-ART

11.2.1. Working Libraries

Many working digital libraries are academic and make information freely available. Some examples are given in the section below describing digital library research. Others are commercial, such as the ACM digital library (http://portal.acm.org/dl.cfm), which contains material from ACM journals, newsletters and conference proceedings. Others, such as BT's digital library which we describe below, are for use within particular organisations. Another category of digital library exists for the explicit purpose of making material freely available. A well-known example of this is Project Gutenberg (http://www.gutenberg.net) which, at the time of writing in autumn 2005, has around 16 000 'eBooks' and claims to be the oldest producer of free e-books on the Internet.

Similarly, the Open Library web site (http://www.openlibrary.org/toc.html) has been created by the Internet Archive, in partnerships with organisations such as Yahoo and HP, to 'demonstrate how books can be represented on-line' and 'create free web access to important book collections from around the world'.

A great deal of digital library software is freely available. One of the best known projects is the Greenstone digital library (http://www.greenstone.org). Available in a wide range of languages, Greenstone is supported by UNESCO and, amongst other applications, is used to disseminate practical information in the developing world. Another example is OpenDLib (http://opendlib.iei.pi.cnr.it), which has been designed to support a distributed digital library, with services anywhere on the Internet.

A recent development from Google sees the world of the public domain search engine encroaching that of the digital library. *Google Scholar* (http://scholar.google.com/) provides access to 'peer-reviewed papers, theses, books, abstracts and other scholarly literature'. It uses the same technology as Google uses to access the public Web and applies this to on-line libraries. This includes using Google's ranking technology to order search results by relevance. In a similar initiative, Yahoo is working with publishers to provide access to digital libraries.

11.2.2. Challenges

> Libraries, museums and archives face huge challenges in the way that they acquire, preserve and offer access to their collections in the digital age. Although having similar objectives, the different types of institution tend to use different technologies and working methods. With more and more digital born documents, new issues are raised in terms of cataloguing, search and preservation. As the different types of institution move closer together, they are seeking common frameworks for managing digital collections and content across the cultural sector.
>
> For users, the value of libraries, museums and archives lies not only in their own resources but as gateways to huge distributed collections in other cultural institutions. This, too, poses major challenges in terms of content management: namely how to provide the user with seamless, high value, interactive services based on these distributed resources. *DigiCULT*

The DigiCULT[1] quote above identifies a central challenge facing digital libraries; that of combining heterogeneity of sources with efficient cataloguing and searching, and with an appearance of seamlessness to the user. The technologies discussed in this book provide a response to this challenge. Through the creation of ontologies and the creation of associated semantic metadata to describe documents, technology can partially automate the process of cataloguing information. Through the use of those ontologies and metadata, our technology will offer an improved search and browse experience. At the same time, work in

---

[1]DigiCULT (http://www.cordis.lu/ist/digicult) is a European Commission activity on Digital Heritage and Cultural Content. It contains within it DELOS and BRICKS, two projects mentioned later in this chapter.

the areas of ontology merging and mapping offers the prospect of seamless access to distributed information.

As long ago as 1995 a workshop held under the auspices of the U.S. Government's Information Infrastructure Technology and Applications Working Group identified five key research areas for digital libraries (Lynch and Garcia-Molina, 1995):

1. *Interoperability*: At one level this is about the interoperability of software and systems. At a deeper level, however, it is about semantic interoperability through the mapping of ontologies. Indeed 'deep semantic interoperability' has been identified as the 'Grand Challenge of Digital Libraries' (Chen 1999).
2. *Description of objects and repositories*: This is the need to establish common schema to enable distributed search and retrieval from disparate sources. Effectively, how can we create an ontology for searching and browsing into which we can map individual library ontologies? Going further, how can we enable individual users to search and browse within the context of their own personal ontologies?
3. *The collection and management of nontextual information*: This includes issues relating to the management, collection and presentation of digital content across multiple generations of hardware and software technologies. Moreover, libraries are now much more than collections of words, but are increasingly rich in audiovisual material, and this raises new research challenges.
4. *User interfaces*: We need better ways to navigate large information collections. One approach is through the use of visualisation techniques. The use of ontologies not only helps navigation but also provides a basis for information display.
5. *Economic, Social and Legal Issues*: These include digital rights management and 'the social context of digital documents'.

Semantic technology makes significant contributions to (1), (2) and (4). Although this book is chiefly concerned with textual material, ontologies can be used to describe the nontextual information referred to in (3). Semantic technology also impacts (5), for example through enhancing knowledge sharing in social groups.

The need for interoperability across heterogeneous data sources is repeated by many authors. A more recent U.S. workshop on research directions in digital libraries identified a number of basic themes for long-term research (NSF, 2003), of which one is interoperability, which it describes as 'the grail of digital libraries research since the early 1990s'. A number of the other themes reiterate the need to overcome heterogeneity.

The NSF workshop also identified 'question answering' as a grand challenge for research, stressing the need to match concepts not just search terms. The use of semantic technology to do just that in a legal application is discussed in Chapter 12 of this book.

11.2.3. The Research Environment

As implied by Lynch and Garcia-Molina (1995), the topic of digital libraries has attracted significant research activity since the 1990s. Some of this work has been with very specific goals. For example, the Alexandria Digital Earth Project (http://www.alexandria.ucsb.edu/), at the University of California, is concerned with geospatial data, whilst other projects have investigated areas such as medical information,[2] music[3] and mathematics.[4]

In the US, an example of the more generic research activities is the Perseus Digital Library project (http://www.perseus.tufts.edu/) at Tufts University. The aim here is 'to bring a wide range of source materials to as large an audience as possible'. The intention is to strengthen the quality of research in the humanities by giving more people access to source material. At the same time, there is a commitment to 'connect more people through the connection of ideas'.

Within Europe, at the beginning of the current decade, the European 5th Framework Programme played a major role in digital library research. One of the 5th Framework Programme projects gave rise to the Renardus service (http://www.renardus.org/) which 'provides integrated search and browse access to records from individual participating subject gateway services'. The subject gateways use subject experts to select quality resources. This overcomes the variability of quality in Web material, although it is admitted that it 'encounters problems with the ever increasing number of resources available on the Internet'. Renardus also enables searching across several gateways simultaneously, based on searching the metadata, not the actual resources.

Another 5th Framework project, Sculpteur[5] (http://www.sculpteur-web.org), used semantic technology for multimedia information management. The target domain is that of museums. An ontology, with associated tools, has been created to describe the objects, whilst a web crawler searches the Web for missing information.

Currently, the 6th Framework Programme is sponsoring significant research in the area of digital libraries. Two major activities initiated at the outset of the Programme are DELOS (http://www.delos.info/) and BRICKS (http://www.brickscommunity.org/). The goal of DELOS is to develop 'the next generation of digital library technologies'. DELOS has seven 'clusters', of which 'Knowledge Extraction and Semantic Interoperability' is one. A key motivation of the cluster is the need for semantic interoperability between the many existing vocabularies within

---

[2]PERSIVAL (http://persival.cs.columbia.edu/) at Columbia University.
[3]VARIATIONS (http://www.dlib.indiana.edu/variations/) at Indiana University.
[4]EULER (http://www.emis.de/projects/EULER/) developed by the Euler consortium as part of the European Community's IST programme.
[5]Semantic and content-based multimedia exploitation for European benefit.

the digital library arena. Other clusters of relevance here are 'Information Access and Personalization' and 'User interface and visualization'. The former is investigating techniques for customising information to suit the characteristics and preferences of users. One goal of the latter is to 'build a theoretical framework from which user interface designers/developers can design digital library user interfaces'. Both are areas where semantic technology has a role to play.

The aim of BRICKS is to 'design and develop an open, user and service-oriented infrastructure to share knowledge and resources in the Culture Heritage Domain' (Meghini and Rissi, 2005). A decentralised peer-to-peer architecture is used to provide access between member institutions of a BRICKS installation. The BRICKS view is that 'semantic interoperability should be considered as an incremental process, starting from 'local agreements'' (Nucci, 2004).

Closely related to digital library initiatives are those in the area of eScience, which are aimed at sharing knowledge and permitting collaboration. Another European 6th Framework project, DILIGENT (http:// diligentproject.org/) is aimed at creating Virtual Dynamic Digital Libraries for virtual eScience communities, using Grid technology. Both BRICKS and DILIGENT have, as goals, interoperability between institutions.

11.3. A CASE STUDY: THE BT DIGITAL LIBRARY

The goal of our digital library case study in SEKT is to investigate how the semantic technologies being developed can enhance the functionality of the digital library. We have seen already that a number of the key challenges facing digital libraries relate to issues of semantics and semantic interoperability. We explain in Subsection 11.3.2 below how semantic technology, and in particular the semantic knowledge access tools described in Chapter 8, are being used to enhance our digital library. Before that, Subsection 11.3.1 briefly describes the BT digital library before the incorporation of semantic technology.

11.3.1. The Starting Point

In developing its digital library case study, SEKT did not start from scratch but from BT's existing digital library. BT began building its digital library in 1994 and over the past decade has developed an online system that offers its users personalisation,[6] linking to full text from abstracts, annotation tools, alerts for new content and the foundations of

---

[6]A personal home page, as discussed shortly.

profiling. A key driver in developing the digital library has been the desire to provide a single interface to the whole collection, drawing together content from a wide variety of publishers.

The BT digital library allows its users to search the library's contents. In addition, they can browse through 'information spaces' that have been created on subjects known to be of interest to people in the company or through the contents of journals in the library. Information Spaces bring together content from the library's databases and details of new books. They are defined by a specific query, that is the documents within the information space are those which would be found by the query. People can 'join' an information space to be alerted to new articles on the subject and can create their own private information spaces for subjects of particular interest to them. Figure 11.1 shows the introductory page to one particular information space. For more information about the use of information spaces in the BT digital library, see Alsmeyer and Owston (1998).

Users can also be alerted when new issues of particular journals are received in the library. Information space membership and journal preferences are used to provide a personalised view of the digital library homepage, as can be seen in Figure 11.2.

The library contains abstracts of all relevant technical papers and the full text of more than a third of all the relevant engineering

Figure 11.1 A digital library information space.

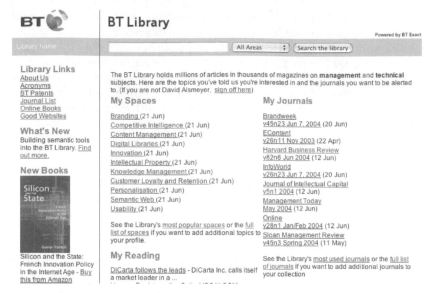

Figure 11.2   A personalised view of the library homepage.

literature—five million articles from over 12 000 publications, including journals, conference proceedings and IEEE Standards. This is provided in the form of two databases, Inspec (http://www.iee.org/Publish/INSPEC/) and ABI/INFORM from ProQuest (http://www.proquest.com/). To be precise, these two databases contain abstracts, each of around 200 words. In some cases the full texts of the articles are available either from the database supplier or from the publisher and the digital library provides links to these.

The library uses software developed by BT's Next Generation Web Research to power its searching and browsing.[7]

11.3.2. Enhancing the Library with Semantic Technology

We look now at ways in which semantic technology is being used to enhance the digital library.

11.3.2.1. Richer Metadata

Both the ABI/INFORM and Inspec abstracts are provided with a significant quantity of metadata, including subject headings from a

---

[7]This software is now available from Corpora software, http://www.corporasoftware.com.

controlled vocabulary, classification codes, publication types etc. The subject headings are taken from hierarchical listings of preferred terms that describe the topics covered by the database. The Inspec thesaurus, for example, contains 9000 preferred terms. As might be expected given Inspec's focus on physics, electronics and computing, the Inspec thesaurus provides a rich set of terms covering these areas and a much higher level of description of more general topics. ABI/INFORM covers the whole breadth of business and management issues and has a vocabulary of about 8000 preferred terms. The schemas used by ABI/ INFORM and Inspec will provide the basis for the initial set of topics to be used to categorise documents semantically, see the discussion of topic hierarchies in Chapter 7. However, these schemas are similar but different and will need to be merged. Moreover, we believe that these schemas will not be adequate to generate a sufficiently fine-grained set of topics. We are taking the set of topics generated from the database schemas and refining this set with ontology learning software developed in SEKT. In addition we will incorporate data already marked up with the RSS standard used for news syndication.

The end result will be a topic hierarchy, that is some topics will be defined as sub-topics of other topics. Thus 'knowledge management' might be a sub-topic of 'information management', and at the same time have 'communities of interest' as its own sub-topic.

It is hoped to add documents from other sources, for example the Web, which do not come with predefined metadata. In this case the associated topics need to be inferred automatically from the text. Addition of material to the library might be prompted by a user finding a page of interest on the intranet or Internet, and wanting to share this with colleagues. We call this 'jotting'. Alternatively, SEKT has developed a focussed crawling facility which, given a seed page, can search the Internet for related pages. Again, these additional pages need to be categorised against the library topics.

11.3.2.2. Enhanced User-Profiling

Information searching, information sharing and the use of information alerts can all be made more precise when we have a profile of the user's interests. Of course, the user himself can create such a profile, using the topic hierarchy. This is similar to the approach taken by Google's personalised web search (http://labs.google.com/personalized). However, this is an overhead from which we wish to relieve the user, at least partially. SEKT is therefore using an automatic user-profiling technique based on analysis of the abstracts downloaded by the user. A typical user may use the Web and intranet at least as much as he uses the digital library. Therefore, to establish his profile, we are taking into account information accessed from the Web and intranet.

11.3.2.3. Unlocking the Documents

Search technology frequently operates at the level of whole documents. A search for a reference to a person will identify documents referring to that individual, but say nothing about where those references occurred in the documents. SEKT intends to unlock the documents by annotating fragments within them, to give a much richer search experience.

As described elsewhere in this book, SEKT will identify 'named entities' in documents. These are entities within a document which can be identified as significant by information extraction techniques. At the simplest level they may be proper nouns, for example names of people or places, which can be identified in English by the use of capital letters. This approach can extend to less tangible entities, for example pieces of legislation. These named entities can then be included as instances in a knowledge base. Typically, within the knowledge base, we distinguish between such learned instances and instances which are 'trusted', that is input into the knowledge base by a domain expert, or acquired from a reliable data source. In the case of learned instances there is the possibility of error. In the case of trusted instances we assume errors will not occur. A specific example of an error would be to associate the text strings 'George Bush' and 'George W Bush' with the same instance in the knowledge base, when the former refers to the famous father and the latter to the famous son. Entities can be identified in the document by a hyperlink, enabling the user to view information about the entity, drawn from the knowledge base. For more detail on this approach, see Kiryakov *et al.* (2004).

11.3.2.4. Enhanced Searching and Browsing

As discussed in detail in Chapter 8, the availability of semantically annotated information resources offers the opportunity to provide more sophisticated search capabilities. Nonetheless, there will always be a need to undertake textual searches, in the traditional way. Indeed, there is a school of thought which says that most users will want to begin any search with a simple text string. At a subsequent stage they will wish to disambiguate between various occurrences of different information entities identified by the same text string.

A typical example might be a company name where that name is used by a variety of entities besides companies or even by several different companies. In the first case the user disambiguates by specifying that he is interested in entities of class 'company'. In the second case he goes further and specifies some characteristic of the company, for example the sector in which it operates. In this case the search is employing the *activeInSector* relationship used to describe companies in our knowledge base. This disambiguation will be achieved through a drop-down menu or menus to provide additional information about the entity sought, for example the class to which it belongs. The design of the user interface is

critical and usability trials will help us understand how people best interact with such semantically enabled knowledge management systems.

The user can also use the topic hierarchy to find relevant documents,
moving up or down the hierarchy to expand or refine the search. For
example, if a search on a particular topic gives no useful hits, the user can
look for a supertopic of the original topic. Conversely, if too many hits
occur, the user can look for sub-topics. Besides searching and browsing
by topic, the user can browse the library using other characteristics of a
document, for example by requesting other documents by the same
author.

Whatever form of search the user employs, the results presented will
be able to take account of the user's profile. For example, the search
string 'visual impairment' will produce quite different results for a user
with an interest in human resources than for a user interested in medical
technology. The user profile will contain a component representing the
longer-term user interests and also a component derived from current
activities, for example the documents read during the current session.
For example, a search for the word 'jaguar' might normally return
information about the South American cat, if the user's profile indicates
an interest in natural history. If our user, on the basis of recently viewed
documents and web pages, appears on this occasion to be interested in
cars, then 'jaguar' could be interpreted differently. On a subsequent day,
the system would revert to its original interpretation, for this particular
user. There may be times when this facility hampers the user, and he or
she must be free to switch off both the long-term and the short-term
aspects of the profile.

One interesting research area is the exploitation of the linkages
between people to influence search results. Assume that two users, A
and B have very similar profiles. User A starts a search with a particular
text string and after some efforts, including using the ontology to achieve
disambiguation, terminates the search with a particular document or set
of documents. Then if user B subsequently searches with the same text
string, the system could reasonably assume that the documents terminat-
ing A's search should be among the most relevant to B.

11.3.2.5. Displaying Results

We are experimenting with improved ways of displaying results to users,
based on an analysis of the semantics.

One approach is to cluster the results according to the principal
themes. A search for 'George Bush' might cluster documents according
to whether they relate to the father or the son. One might go further and
categorise documents according to some aspect of George Bush, for
example foreign policy, domestic policy etc.

An analysis of semantics can also be used to better précis documents.
In the limit, instead of a list of documents one could present the user

with simply the required information, drawn from all the relevant documents.

Experimentation is necessary to understand what the user wants. Underlying whatever techniques are used must be the insight of Herbert Simon, re-quoted in (NSF, 2003): 'information ... consumes the attention of its recipients. Hence a wealth of information creates a poverty of attention, and a need to allocate that attention efficiently among the overabundance of information sources that might consume it'. Our first priority is always to ensure that the right information is presented to the user. The next priority must be to ensure that it is done in a way which minimises the consumption of that user's attention.

11.3.2.6. Connecting Ideas, Connecting People

In an earlier section, we noted that the Perseus Digital Library project had a commitment to 'connect more people through the connection of ideas'. This statement embodies the idea that a digital library should be a community of people as well as a collection of documents. By understanding usage patterns at the semantic level, semantic technology can identify experts as well as communities of interest. In the former case these experts will hopefully be available to give advice, although possibly through the intermediary of the digital library to provide them with anonymity. In the latter case, we hope to help create communities of mutually supportive users with common interests.

11.4. THE USERS' VIEW

The development of any system should be guided by a comprehensive view of what users actually want from that system. At the same time, asking users about their requirements is notoriously difficult. When the proposed system includes radically new functionality with which the user is not familiar, users may expect too much from the technology. Frequently, however, they expect too little and ask for more of the same, but simply faster and cheaper. Even when the potential of a new technology is described, there is a difference between imagining the possibilities and actually using the resultant system. Hence any system design using previously untried technology must take into account what users say they want, but at the same time not let user feedback close off any avenues which use the technology in radically new ways.

Within the digital library case study in the SEKT project a number of methods were employed to obtain a comprehensive view of what users want from digital libraries, so that these requirements could be interpreted in terms of the capabilities of semantic technology.

Initially a questionnaire and focus group were used to gauge user requirements. Much that was learned was very generic and did not relate particularly to the capabilities of semantic technology. However, we did learn that our users wanted improved ways of searching, including the ability to search on attributes of a document; and that they wanted searches to take into account their profile of interests.

The next stage was a questionnaire which asked specific questions about search functionality. Just under 90 people responded to this questionnaire. The responses revealed considerable enthusiasm for a facility which summarised a set of results. A search function which took into account personal preferences was also popular, as was attribute-based search. Also well up the list of requirements was the ability, if a particular article was not available in the digital library, to search it out on the Web.

Amongst the other popular features were:

- A search function which suggests candidate topic areas in which to search.
- The ability to enter natural language queries.
- The highlighting of named entities, for example people and companies, and access to further information about those entities.

Less popular was an application to perform regular searches motivated by personal information, for example held in the user's calendar. The majority of people ranked this 'useful', but only a very small number ranked it 'very useful', suggesting that enthusiasm for such proactive systems is lukewarm.

Our final technique for understanding users' requirements was that of user preference analysis. The essence of this is to investigate the trade-off, from the user's perspective, between various proposed enhancements. One comparison was between precision and recall. Semantic technology can in principle improve both. Precision is enhanced by the ability to specify the nature of the entity being sought, for example that the string 'BT' corresponds to a company in the telecommunications sector. Recall is enhanced by the ability to understand that different text strings represent the same entity, for example the 'George Bush', 'George W Bush' and 'The President' all describe the same person. Nevertheless, there is some trade-off between the two capabilities. Systems which are too keen to identify equivalences, in the interests of recall, may do so at the risk of creating false equivalences and damaging precision. A sample of users was asked to rate their preference between these two capabilities. When the results were analysed, there appeared to be two clusters of users: one with a clear preference for precision and another where the users gave equal weight to precision and recall.

These studies should only be taken as a guide. When users are confronted with real systems they are likely to react differently than when confronted with questionnaires and in focus groups. However, they offer a starting point for understanding users, to be taken into account when faced with trade-offs in designing systems, and to be further tested by users reaction to real semantic digital library systems.

11.5. IMPLEMENTING SEMANTIC TECHNOLOGY IN A DIGITAL LIBRARY

11.5.1. Ontology Engineering

A well-designed ontology is essential for a successful semantic application. Within SEKT we are adopting a layered approach. In the lower layers we have a general ontology, which we call Proton (PROTo Ontology, http://proton.semanticweb.org). The classes in this ontology are a mixture of very general, for example Person, Role, Topic, Time-Interval and classes which are more specific to the world of business, for example Company, PublicCompany, MediaCompany. See Chapter 7 for more detail.

Above this we have the PROTON Knowledge Management ontology, which contains classes relating to knowledge management. Examples are UserProfile and Device.

Finally, each of our three case studies has its own domain-specific ontology. In the case of the digital library, this will contain classes relating to the specifics of the library, for example to the particular information sources available.

A strength of an approach based on the use of an ontology language such as OWL, is the ability to accommodate distributed ontology creation activities, for example through defining equivalences. Nonetheless, where possible the creation of duplicate ontological classes should be avoided and where appropriate we make use of existing well-established ontologies, for example Dublin Core.[8]

Mention has been made of the use of a topic hierarchy. Within PROTON there is a class, 'Topic'. Each individual topic is an instance of this class. However, frequently a topic will be a sub-topic of another topic, for example in the sense that a document 'about' the former should also be regarded as being about the latter. Since topics are instances, not classes, we cannot use the inbuilt subclass property, but must define a new property *subTopic*. Such a relationship must be defined to be transitive, in the sense that if A is a sub-topic of B and B is a sub-topic of C, then A is also a sub-topic of C.

This approach, based on defining topics as instances and using a subTopic property rather than defining topics as classes and using the sub-class relation, is chosen to avoid problems in computational tractability. In particular, this enables us to stay within OWL DL. It follows

---

[8]http://dublincore.org/

approach 3 in Noy (2005). Again, for a more detailed discussion, see Chapter 7.

11.5.2. BT Digital Library End-user Applications

The following end user applications are available:

  (i) a semantic search and browse application,
 (ii) a knowledge sharing application,
(iii) a personal search agent,
(iv) semantically enabled information spaces.

All applications were built upon the core technologies of ontology creation; named entity identification and annotation; ontology maintenance and ontology mediation.

The semantic search and browse application combines free-text search with a capability to query over the ontology and knowledgebase as described in more detail in Chapter 8. The search and browse application augments the more traditional practice of presenting the results of a query as a ranked list of documents with an approach where knowledge contained within documents is presented in a more meaningful way to the user. Named entities, for example company names, are identified and relevant supplementary information is presented to the user. In addition, user-specific, interest-based profiles are constructed in accordance with a user's interaction with the digital library and other WWW and intranet information sources, giving an element of context to the user's search.

The semantic knowledge sharing application enables users to annotate digital library documents, WWW or Intranet pages with topics selected (semi-automatically) from the digital library topic ontology, to share that information with colleagues, and to recall annotated pages at a later date more easily. Our user can also add a comment, for subsequent viewing by his colleagues. The essence of our approach is that sharing is not achieved by pushing information to colleagues, for example via email. Instead, web-pages marked by a user as being of particular interest or value, are presented prominently when they occur amongst the search results of that user's colleague, or when he or she comes across them in browsing. The incentive to share arises from the fact that the sharing mechanism is exactly that of bookmarking, that is in bookmarking the page for himself, the user is sharing it with colleagues.

The personalised semantic search agent collects relevant content from the digital library and WWW on behalf of a user, and gives improved relevance and timeliness of the delivery of information. Named entities within the search agent's results are highlighted. The approach builds on that of KIM, see Chapter 7.

In the original digital library, information spaces were defined by a search, and this remains the case in the semantically enhanced library. The difference is that the defining search may now be semantic instead of textual, or even a combination of semantic and textual.

11.5.3. The BT Digital Library Architecture

The BT digital library is based on a 5-layer architecture comprising the persistence layer, the semantic layer, the integration layer, the application layer and the presentation layer. Access to the applications is provided by a BT digital library semantic portal. The majority of users access the BT digital library applications from a desktop or laptop PC. Some mobile users require access to business critical information, for example relevant breaking news updates, from handheld or PDA devices. The user interfaces to the applications are presented according to the capabilities of the device being used and any preferences set by the user. Note that this architecture, which is illustrated in Figure 11.3, provides the user functionality at 'run-time'. A separate set of functions are used at 'ontology engineering time', for example for creating and editing ontologies and for creating mappings between ontologies.

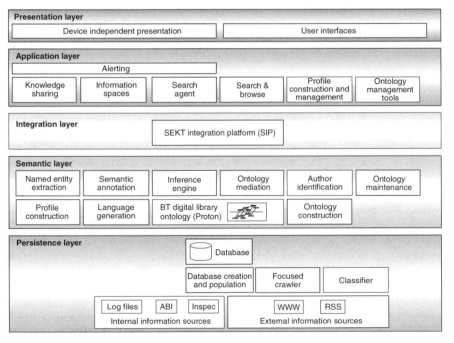

Figure 11.3 The BT digital library run-time architecture.

11.5.3.1. The Persistence Layer

The persistence layer comprises the internal sources of information, for example the subscribed ABI and Inspec databases, and external sources of information, for example RSS items. The SEKT components that draw together relevant content for the digital library, for example the focused crawler and the components that populate the database and build profiles from an analysis of the log files are incorporated into the persistence layer. The Inspec and ABI records, RSS items, and the text extracted from web pages and RSS items are stored together with their associated metadata in the database. A classifier classifies the web pages and RSS items against topics in the BT digital library ontology.

11.5.3.2. The Semantic Layer

The semantic layer is concerned with the creation, enhancement, maintenance, and querying of ontological information that is linked to the data stored in the persistence layer.

Metadata associated with Inspec, ABI and RSS items is transformed into BT digital library ontology-specific metadata. Where possible the original data is enhanced with metadata that is created from or identified within the data itself, for example named entities such as the name of a company can detected in the abstract of a ABI record.

The BT digital library ontology is based on the PROTON general ontology, as already described. This defines the top-level generic concepts required for semantic annotation, indexing and retrieval, e.g. concepts such as author and document. This base ontology is extended with some additional classes and properties that are required to facilitate the SEKT-specific and case study-specific applications and functions.

User interest profiles, which are also stored in the ontology, are constructed from an analysis of user interaction with the BT digital library (from the digital library Web server log files) and from the content of the Web pages that a user accesses. Software within the user's Web browser analyses documents accessed (for example, treating them as 'bags of words') and creates a vector representing the user's interests. These vectors are mapped to the most relevant topics in the BT digital library ontology. In turn, the topics are then added to the user's profile under the control of the user.

The ontology store includes not just the PROTON ontology but also a set of rules to be run when a query is executed. These rules can be used to enable sophisticated query facilities, and also to enable a mapping between the ontologies.

Components in the semantic layer augment the ABI, Inspec and Web data with supplementary metadata. The named entity identification and annotation components identify named entities such as people's names, place names, and company names within the library content, and provide the semantic annotations which can be queried by the semantic query component.

The ontology construction components create the fine-grained sub-topic structure within a set of documents (textual items) classified by an information space. The ontology construction components also enable new information to be classified against topics in the BT digital library ontology.

Instance disambiguation components identify potential ambiguities in the instance data, for example the author identification component identifies equivalent author names within the BT digital library ontology and disambiguates where authors share a common name and initials. This in turn enables further metadata to be generated that links instances concerned with a particular author.

The natural language generation component enables natural language statements to be built from the information held in the ontology. Such statements are used to enhance the way in which information is presented to users. For example information about people, companies, related topics and relevant information spaces is presented to the user in preference to listing a set of search results. Additionally, natural language generation can be used to generate descriptions of topics and information spaces.

The components that are required to populate, annotate, store, index and manage the BT digital library ontology and enable the ontology to evolve over time are provided in the semantic layer. The process of adapting the ontology is supported by components that discover changes in the underlying data and that can adapt the ontology incrementally in accordance with those changes. End user interaction with the digital library is also analysed to enable changes to be made to the ontology that would best suit the needs of end users.

The ontology mediation component unifies any underlying ontologies that are used in the BT digital library, for example ontology-mapping rules enable equivalent classes in different underlying ontologies to be mapped to each other, thereby facilitating querying across equivalent classes.

11.5.3.3 The Integration Layer

The integration layer provides the infrastructure that enables the applications to be built from SEKT components (in the semantic layer). The integration functions are provided by SEKT Integration Platform (SIP). The SIP infrastructure also enables semantic layer components to be integrated, for example the integration of data mining components with GATE.[9]

---

[9]http://gate.ac.uk/

11.5.3.4. The Applications Layer

The BT digital library applications utilise the components of the semantic layer. In general, applications such as the search and browse, and, search agent applications, query the data held in the BT digital library ontology through the inference engine via the SIP. The architecture also allows for applications to interface directly to semantic layer components where necessary. The alerting component, which is common to all applications that push information to users, enables information alerts to be delivered at a time and in a format that is suitable to the user. A profile construction component, which is integrated with a web browser, enables profiles of users' interests to be constructed.

11.5.3.5. The Presentation Layer

Client devices interact with the presentation layer of the architecture. A device independent presentation component presents the user interface for each end-user application according to the capabilities of the device being used and to the preferences set by the user.

11.5.4. Deployment View of the BT DIGITAL LIBRARY

The BT digital library architecture has been implemented on two Sun Microsystems servers. All components in the semantic, application and presentation layers have been deployed on a Sun Blade 1500 server running SunOS 5.9. The back end databases for Inspec and ABI/ INFORM are provided on the existing BT digital library Sun Fire V240 server, running SunOS 5.8.

11.6. FUTURE DIRECTIONS

Today digital libraries are walled gardens; stocked with knowledge of known provenance and hence in which a degree of trust is possible; relatively well catalogued and provided with metadata; and for which a charge exists for entry. Outside these walls lies the Web with a vast quantity of information; some of it immensely valuable but much of dubious provenance and validity; with limited or no cataloguing and limited metadata; but free for all.

The history of information and communication technologies is one of disappearing barriers. Witness the attempt to create walled gardens by companies such as AOL in the previous decade. Digital libraries will not escape this trend.

The future Semantic Web will include a wide variety of heterogeneous resources. de Roure *et al.* (2005) describe a Semantic Grid which effectively subsumes the Semantic Web and includes resources ranging from powerful computational resources to sensor networks. Amongst these will be the components of a digital library. Yet the digital library as an identifiable entity may have ceased to exist. Instead the user of the Web will see a network of resources, of varying provenance, trustworthiness and cost. Much will be free, but where payment is justifiable, then it will be required. The walled garden will have ceased to exist, but instead individual items within the whole landscape will have controlled access.

The resources themselves will vary enormously. Not just text and multimedia in the conventional sense, but software and data objects of all sorts. The last of these will include the results of scientific experiments, so that researchers will not just read their colleagues research results on-line, but also have access to the raw data and be able to repeat the analyses. They will have access to some data even as it is being created, for example sensor data.

All this data will be linked. A paper on the Web will link to its references. The paper will also be linked to the data used to generate the published results. Data in a databank will link to the papers which have made use of it.

There will be an enormous richness of metadata. For example, we are used today to seeing the finished product of an intellectual process; for example the scientific paper which creates new ground-breaking insight. How much could we learn from understanding the process which created it; for example the reasons why a particular approach is used, and why so many others are rejected. All this information can be captured as the intellectual process itself is taking place, and treated as metadata.

The suggestion has even been made that the paper, as a linear narrative, may lose its monopoly as a medium of communication, at least in the scientific world (de Waard, A 2005). Perhaps to be complemented by 'sets of triples, or at least annotated hypertext'. More prosaically one could imagine authors plagiarising their own, or even others work, by hyperlinking sections from previous work into new work, for example to provide a background to the new work.

To exploit its full benefits, new technology demands new ways of working. The introduction of information technology should always be accompanied by a redesign of business processes. One author has forcibly made the point that digital libraries must support new ways of intellectual work (Soergel, 2002). So our technology must be seamlessly integrated into the systems which support a user's work; and we must seek to go beyond the limitations of our paper-based metaphors and truly exploit the power of the technology.

To achieve all this, significant research is still needed. Just as in other chapters' authors have stressed the need for more research into the core semantic technologies, so here we stress the need for more research into exploiting those technologies to create the digital libraries of the future.

Encompassed within this research will be work to understand how the new ways of organising knowledge enable and demand new ways of performing knowledge work; so that the new technology can radically enhance our intellectual activity.

REFERENCES

Alsmeyer D, Owston F. 1998. Collaboration in Information Space. *Proceedings of Online Information 98*, Learned Information Europe, Ltd, pp 31–37.

Chen H. 1999. Semantic Research for Digital Libraries, *D-Lib Magazine*, Vol. 5, No. 10, October 1999. http://www.dlib.org/dlib/october99/chen/10chen.html

de Roure D, *et al.* 2005. The Semantic Grid: Past, Present and Future. *Proceedings of the IEEE* 93(3), pp 669–681.

de Waard A. 2005. Science Publishing and the Semantic Web. In *Industry Forum: Business Applications of Semantic Web Challenge Research*, at 2nd European Semantic Web Conference 2005.

Kiryakov A, Popov B, Terziev I, Manov D, Ognyanoff. 2004. Semantic annotation, indexing, and retrieval. *Journal of Web Semantics* 2:49–79.

Lynch C, Garcia-Molina H. 1995. Interoperability, Scaling and the Digital Libraries Research Agenda. A report on the May 18–19th 1995 IITA digital libraries workshop. http://dbpubs.stanford.edu:8091/diglib/pub/reports/iita-dlw/main.html

Meghini C, Risse T. 2005. BRICKS: A Digital Library Management System for Cultural Heritage. In ERCIM News, No. 61, April 2005, http://www.ercim.org/publication/Ercim_News/enw61/meghini.html

Noy N. 2005. Representing Classes as Property Values on the Semantic Web, W3C Working Group Note 5th April 2005, http://www.w3.org/TR/2005/NOTE-swbp-classes-as-values-20050405/

NSF. 2003. Knowledge Lost in Information, Report of the NSF Workshop on Research Directions in Digital Libraries, June 15–17, 2003. http://www.sis.pitt.edu/~dlwkshop/report.pdf

Nucci F. 2004. BRICKS Ontology Approach 'Emergent Semantics', http://www.w3c.it/events/minerva20040706/nucci-en.pdf

Soergel D. 2002. A Framework for Digital Library Research. in *D-Lib Magazine*, December 2002, Vol. 8, No. 12, http://www.dlib.org/dlib/december02/soergel/12soergel.html

12

Semantic Web: A Legal Case Study

Pompeu Casanovas, Núria Casellas, Joan-Josep Vallbé,
Marta Poblet, V. Richard Benjamins, Mercedes Blázquez,
Raúl Peña and Jesús Contreras

12.1. INTRODUCTION

Socio-legal studies have used the notion of 'legal culture' in many senses
since Friedman initially coined the term as 'the network of values and
attitudes related to law' (Friedman, 1969) and further distinguished
between the 'external legal culture'—the culture of the general population—and the 'internal culture'—'the legal culture of those members of
society who perform specialized legal tasks' (Friedman, 1975).

Notwithstanding the valuable contribution of the concept to the
analysis of legal systems, criticisms were made because of its lack of
measurability. In this regard, Blankenburg proposed to split the concept
into various levels and variables of analysis, namely: (i) the ideas and
expectations of justice; (ii) the doctrine of major families of legal systems;
(iii) legal training, legal professions, courts, and their procedures; (iv) the
way legal institutions actually work, and (v) the degree of trust of people
in them (Blankenburg, 1999).

However, we have argued elsewhere that the problem of linking this
general institutional framework of legal behavior with the more concrete
procedures of thinking, deciding, and ruling still remains unsolved
(Casanovas, 1999). The work described here is an attempt to identify,
organize, model, and use the practical knowledge produced by judges in
judicial settings. We will refer to 'judicial culture' or, more specifically, to

Semantic Web Technologies: Trends and Research in Ontology-based Systems
John Davies, Rudi Studer, Paul Warren © 2006 John Wiley & Sons, Ltd

'judicial knowledge' to describe the whole range of cognitive skills and technical resources displayed by judges in judicial units to think, decide, and judge.

This chapter describes the different steps taken in the legal case study towards the design and development of the Iuriservice system. Iuriservice is a web-based application that retrieves answers to questions raised by incoming judges in the Spanish judicial domain. Iuriservice provides these newly recruited judges with access to frequently asked questions (FAQ) through a natural language interface. The judge describes the problem at hand and the application responds with a list of relevant question-answer pairs that offer solutions to the problem faced by the judge altogether with a list of relevant judgments. This application can also be used as a traditional FAQ system, by selecting the appropriate question from a list. In this way, Iuriservice aims at organizing, modeling, and making judicial knowledge usable to any incoming judge.

12.2. PROFILE OF THE USERS

Identifying the problems that newly recruited judges face in daily work and modeling judicial knowledge are basic purposes of the legal case study. To fulfill those objectives, extended fieldwork was performed from March to September 2004.[1] The research targeted the judges of the 52nd class of the Judicial School, who filled vacancies in first instance courts scattered throughout Spain (14 of 17 Autonomous Communities were visited). This group of judges took office by early 2002, so that they had already spent 2 years in office. Consequently, the 52nd class fulfilled our two basic ethnographic requirements: they were newly recruited judges who, at the same time, had spent time enough in office so as to provide researchers with a number of questions regarding daily problems, on-duty periods, and legal procedures at large.

Interviews with newly recruited judges contain a number of variables relevant to describe the organizational context of users (i.e., work conditions, organization of judicial units, professional contacts, etc.). The fieldwork also aimed at obtaining an accurate profile of judges as

---

[1]The UAB Observatory of Judicial Culture (OJC) had already conducted a national survey on newly recruited judges in 2002 (Ayuso *et al.*, 2003). The survey consisted of in depth interviews to 130 incoming judges. Interviews were conducted by their own peers, still at the Judicial School, as part of their training. Judges were taught how to perform the interviews so that they could also obtain information about what they could expect to encounter in their future workplaces. To compare results, 141 senior magistrates were also interviewed.

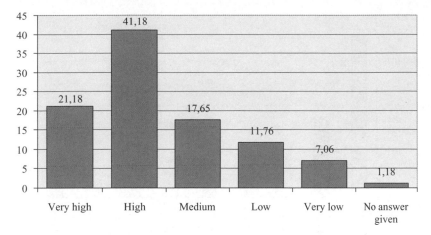

Figure 12.1   Perception of judges of work pressure (2004).

potential users of Iuriservice. Results therefore concentrate on both sociological variables and IT skills (use of Internet, use of hardware and software applications, use of legal databases, etc.).

As regards organizational contexts of users, results show that most newly-recruited judges work under time pressure. Almost 95 % of judges interviewed declared to bring work home in the evening and 87 % added that they worked over the weekends as well. On average, judges work 24 extra hours per week and 63 % of them consider that work pressure is 'high' or 'very high' (see Figure 12.1)

With respect to IT skills, although judges typically argue in interviews that they have no time no navigate through the Internet, results indicate the growing use of the Internet among them (only 19 % of them declare not using it). The page of the Official Bulletin of the State is the most accessed site (45 % of cases), followed by legal information sites in general (20 %).

To the question of 'which would you like to find if judges were given a web service system' the majority of them proposed a site where doubts regarding professional cases could be shared and discussed (see Figure 12.2).

Nevertheless, results also reveal that, despite growing use of the Internet, users of the system will be judges who have medium or low technological abilities, not fully acquainted to new technologies. At the same time, they are willing to accept them, provided they facilitate decision-making and management of daily caseload. The main conclusion relevant to the design of Iuriservice, therefore, is that the web-based platform should be easy to learn and user-friendly for judges.

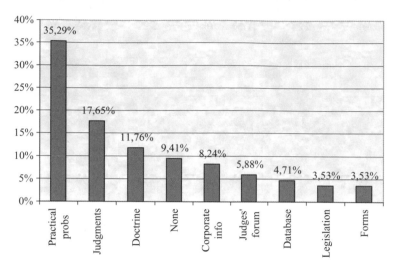

Figure 12.2 Preferences of judges regarding potential web services (2004).

12.3. ONTOLOGIES FOR LEGAL KNOWLEDGE

Legal ontologies are different from other domain ontologies in two ways. On the one hand, although legal statutes, legal judgments, or jurisprudence are written both in natural and technical language, all the common sense notions and connections among them, which people use in their everyday life, are embodied in the legal domain.

On the other hand, the strategy of ontology building must take into account the particular model of law that has been chosen. This occurs in a middle-out level that it is possible to skip in other ontologies based in a more contextual or physical environment.

Therefore, the modeling process in the legal field usually requires an intermediate level in which several concepts are implicitly or explicitly related to a set of decisions about the nature of law, the kind of language used to represent legal knowledge, and the specific legal structure covered by the ontology. There is an interpretative level that is commonly linked to general theories of law. This intermediate level is a well-known layer between the upper top and the domain-specific ontologies, especially in 'practical ontologies.'[2] We may also implicitly find this distinction between an ontology layer and an application layer in

---

[2]An interpretation is the mapping (semantics) from one application instance (conceptual schema) syntactically described in some language into the ontology base, which is assumed to contain conceptualizations of all relevant elementary facts [· · ·]. The interpretation layer constitutes an intermediate level of abstraction through which ontology-based applications map their syntactical specification into an implementation of ontology "semantics" (Jarrar and Meersman, 2001).

cognitive modeling, in which categories, concepts and instances are distinguished.[3] But the most striking feature of the legal ontologies constructed so far is that the intermediate layer is explicitly occupied by a kind of high conceptual constructs provided by general theories of law instead of empirical or cognitive findings.

12.3.1. Legal Ontologies: State of the Art

At present, many legal ontologies have been built. One current way of describing the actual state of the art is to identify the main current legal ontologies (Visser and Bench-Capon, 1998; Gangemi and Breuker, 2002; Rodrigo *et al.*, 2004; Casanovas *et al.*, 2005b):

- LLD [Language for Legal Discourse: (McCarty, 1989)], based on atomic formula, rules, and modalities;
- NOR [Norma: (Stamper, 1996)] based on agents behavioral invariants and realizations;
- LFU [Functional Ontology for Law: (Valente, 1995)] based on normative knowledge, world knowledge, responsibility knowledge, reactive knowledge, and creative knowledge;
- FBO [Frame-Based Ontology of Law, (van Kralingen, 1995; Visser 1995)], based on norms, acts, and descriptions of concepts;
- LRI-Core Legal Ontology (Breuker *et al.*, 2002), based on objects, processes, physical entities, mental entities, agents, and communicative acts;
- IKF-IF-LEX Ontology for Norm Comparison (Gangemi *et al.*, 2001), based on agents, institutive norms, instrumental provisions, regulative norms, open-textured legal notion, and norm dynamics.

At the moment, at least thirteen different legal ontologies have been identified (see Figure 12.3 below), corresponding to 10 years of research. A. Valente (2005) has recently provided the following account of their stage of development, adding to the classical ones recent work made by Mommers, Lame, Leary, Vanderberghe, Zeleznikow, Saias, and Quaresma Ha, etc.[4]

The legal ontologies described above have been built up with several purposes: information retrieval, statute retrieval, normative linking,

---

[3]'Cognitive informatics is the study of the cognitive structure, behavior, and interactions of both natural and artificial computational systems, and emphasizes both perceptual and information processing aspects of cognition [···]. Constructing the mental model of human expertise within the context of a particular problem-solving task is referred to as cognitive or conceptual modeling [···]. An ontology can also be regarded as a description of the most useful, or at least most well trodden organization of knowledge in a given domain' (Chan, 2003: 269–270).

[4]At present, there are even more ontological attempts with respect to particualr domains of law, for example intellectual property rights (Gil *et al.*, 2005).

Ontology or Project	Application	Type	Role	Character
McCarty's Language of Legal Discourse	General language for expressing legal knowledge	Knowledge representation, highly structured	Understand a domain	General
Valente & Breuker's Functional Ontology of Law	General architecture for legal problem solving	Knowledge base in Ontolingua, highly structured	Understand a domain, reasoning and problem solving	General
Van Kralingen & Visser's Frame Ontology	General language for expressing legal knowledge, legal KBSs	Knowledge representation, moderately structured (also as a knowledge base in Ontolingua)	Understand a domain	General
Mommer's Knowledge-based Model of Law	General language for expressing legal knowledge	Knowledge base in English very highly structured	Understand a domain	General
Breuker & Hoekstra's LRI-Core Ontology	Support knowledge acquisition for legal domain ontologies	Knowledge base in DAML+OIL/RDF using Protege (converted in OWL)	Understand a domain	General
Benjamins, Casanovas et al.'s ontologies of professional legal knowledge (OPJK)	Intelligent FAQ system (information retrieval) for judges	Knowledge base in Protégé, moderately structured	Semantic indexing and search	Domain
Lame's ontologies of French Codes	Legal information retrieval	NLP oriented (lexical), knowledge base, lexical, lightly structured	Semantic indexing and search	Domain
Leary, Vanderverghe & Zeleznikow's Financial Fraud Ontology	Ontology for representing financial fraud cases	Knowledge base (schema) in UML, lightly structured	Semantic indexing and search	Domain
Gangemi, Sagri & Tiscornia's JurWordNet	Extension to the legal domain of WordNet	Lexical Knowledge base in DOLCE (DAML), lightly structured	Organize and structure information	General
Asaro et al.'s Italian Crime Ontology	Schema for representing crimes in Italian law	Knowledge base (schema) in UML, lightly structured	Organize and structure information	Domain
Boer, Hoekstra & Winkel's CLIME Ontology	Legal advice system for maritime law	Knowledge base in Protégé and RDF, moderately structured	Reasoning and problem solving	Domain
Lehman, Breuker & Browver's Legal Causation Ontology	Representation of causality in the legal domain	Knowledge base lightly structured	Understand a domain	Domain
Delgado et al's IPROnto (Intellectual Property Rights Ontology)	Integrating XML DTDs and Schemas that define Rights Expression Languages and Rights Data Dictionaries	Knowledge base: first version in DAML+OIL (2001), current version OWL (2003)	Interoperability between Digital Rights Management (DRM) systems	Domain

Figure 12.3 A. Valente (2005: 72) (updated and reproduced with permission).

knowledge management, or legal reasoning. Although the legal domain remains very sensitive to the features of particular statutes and regulations, some of the Legal-Core Ontologies (LCO) are intended to share a common kernel of legal notions. LCO remain in the domain of a general knowledge shared by legal theorists, national, or international jurists and comparative lawyers.

However, our data indicate that there is a kind of specific legal knowledge, which belongs properly to the legal and judicial culture, and that is not being captured by the current LCO.

12.3.2. Ontologies of Professional Knowledge: OPJK

Professional knowledge is a specific type of knowledge related to particular tasks, symbolisms, and activities possessed by professionals which enable them to perform their work with quality (Eraut, 1992). Professional knowledge, then, includes propositional knowledge (knowing that), procedural knowledge (knowing how), personal knowledge (intuitive, pre-propositional), and principles related to morals or deontological codes.

Judges, prosecutors, and other court staff share only a portion of the legal knowledge (mostly, the legal language and the general knowledge of statutes and previous judgments). But there is another part of this legal knowledge, the knowledge related to personal behavior, practical rules, corporate beliefs, effect reckoning, and perspective on similar cases, that remains implicit and tacit within the relation among judges, prosecutors, attorneys, and lawyers.

Consider the following problem, extracted from different kinds of transcriptions of the research protocols, contained in Figure 12.4 below:

Technically speaking, these problems are not complex. However, they are difficult to solve. The judges' original question cannot be answered by simply pointing out a particular statute or legal doctrine. This is not only an issue of normative information retrieval. What is at stake here is a different kind of legal knowledge, a professional legal knowledge (PLK) (Benjamins *et al.*, 2004). What judges really seek are some clues,

---

"I have the following problem, let us see if you come up with something: one woman files a suit (she went to hospital to get care for the bruises) but then she forgives her husband, tells us that they both were drunk that night but are very happy (to show us how happy they are she even insists on remaining in the room while he gives a statement). She keeps saying no way, she is not going to denounce her husband, and she has forgiven him.

Since it's a public offence I go ahead and then the prosecutor [*fiscala* [fem.]] gets angry with me because she appoints him to court [*lo persona*] and wants me to appoint her wife to instruct her on her rights [*instruirle de sus derechos*].

The issue has no objective criminal entity [*entidad penal objetiva*]; to criminalize those little things seems to me really nonsense, it may even be worse regardless of the prosecutor moving forward." [May 2004, personal communication]

Figure 12.4 Literal transcription of a practical procedural problem on gender violence. Pompeu Casanovas. (personal e-mail communication, May 2004, reproduced with the permission of the sender.)

some hints or well-grounded practical guidelines that refer to the problem they have before them when they put the question or start the query.

In this regard, the design of legal ontologies requires not only to represent the legal, normative language of written documents (decisions, judgments, rulings, partitions...), but also the professional knowledge sorted out from the daily practice at courts.

From this point of view, professional knowledge of a legal topic (such as e.g., gender violence) involves a particular knowledge of: (i) statutes, codes, and legal rules; (ii) professional training; (iii) legal procedures; (iv) public policies; (v) everyday routinely cases; (vi) practical situations; (vii) people's most common reactions to previous decisions on similar subjects.

This Professional Legal Knowledge (PLK) is: (i) shared among members of a professional group (e.g., judges, attorneys, prosecutors...); (ii) learned and conveyed formally or most often informally in specific settings (e.g., the Judicial School, professional associations—the Bar, the Judiciary, etc.); (iii) expressible through a mixture of natural and technical language (legalese, legal slang); (iv) nonequally distributed among the professional group; (v) nonhomogeneous (elaborated on individual bases); (vi) universally comprehensible by the members of the profession (there is a sort of implicit identification principle).

Professional knowledge is then a context-sensitive knowledge, anchored in courses of action or practical ways of behaving. In this sense, it implies: (i) the ability to discriminate among related but different situations; (ii) the practical attitude or disposition to rule, judge, or make a decision; (iii) the ability to relate new and past experiences of cases; (iv) the ability to share and discuss these experiences with the peer group.

12.3.2.1. Ontologies of Professional Legal Knowledge

In order to build Ontologies of Professional Legal Knowledge (OPLK) we believe that we have to take into account the kind of situated knowledge that judges put into practice when they store, retrieve, and use PLK to make their most common decisions.[5]

On the one hand, for all practical purposes there is no such thing as absolute meaning: everything must ultimately be the result of agreements among human agents such as designers, domain experts, and

---

[5]We use 'situated knowledge' in a similar way in which Clancey et al. (1998: 836) and Menzies and Clancey (1998: 767–768) talk about 'situated cognition:' the concrete use of knowledge which is partially shared and unequally distributed through a certain 'community of practice' which is able to use and reuse this same knowledge while transforming it. Other related concepts close to 'situated knowledge' are the ideas of 'situated communities,' 'situated meaning,' 'organizational memory,' and 'corporate ontologies.'

users (Jarrar and Meersman, 2001: 3). On the other hand, in ontology knowledge modeling a concept is neither a class nor a set: the concepts which represent the term's meaning are structured into binary trees based on couples of opposite differences (Roche, 2000: 188).

Ontologies of PLK model the situated knowledge of professionals at work. In our particular case we have before us a particular subset of PLK belonging specifically to the judicial field. Therefore, we will use the term Ontology of Professional Judicial Knowledge (OPJK) to describe our conceptual specifications of the knowledge contained in our empirical data.

12.3.2.2. Ontology of Professional Judicial Knowledge (OPJK)

The OPJK is learnt from of the competency questions posed by the judges during their interviews. Modeling this professional judicial knowledge required the description of this knowledge, as it was perceived by the judge.

The OPJK has, currently, 700 terms, mostly relations and instances as a result of a choice to minimize the concepts at the class level when possible. Some top classes of the domain ontology identified are: *CalificacionJurdica* [LegalType], *Jurisdiccion* [Jurisdiction], *Sancion* [Sanction], *Acto* [Act], (which includes as subclasses *ActoJurdico* (LegalAct), *Fase* [Phase], and *Proceso* [Process]). These latter classes contain those taxonomies and relations related to the different types of judicial procedures (both, criminal and civil, or private) and the different stages that these procedures may have (period of proof, conclusions, appeal, etc.). The introduction of the concept *Rol* [Role] allowed for the specification of different situations where the same agent could play different parts. In the case of OPJK, the class *Rol* contains the concepts and instances of procedural roles [*RolProcesal*] that an agent might play during a given judicial procedure.

Some of the properties/attributes of concepts and relations between concepts are, for example, that *Agente* has_role, is_involved_in_facts, that *ActoProcesal* has_document, that *FaseProcesal* begins_with, ends_with, is_followed_by, that *ProcesoJudicial* has_phase, and that *RolProcesal* is_played_by (Figure 12.5).

12.3.3. Benefits of Semantic Technology and Methodology

12.3.3.1. Ontology Learning

The TermExtraction feature of TextToOnto[6] provided, together with another textual statistics programe (Alceste),[7] a good basis for

---

[6]http://kaon.semanticweb.org/
[7]http://www.image.cict.fr/index_alceste.htm

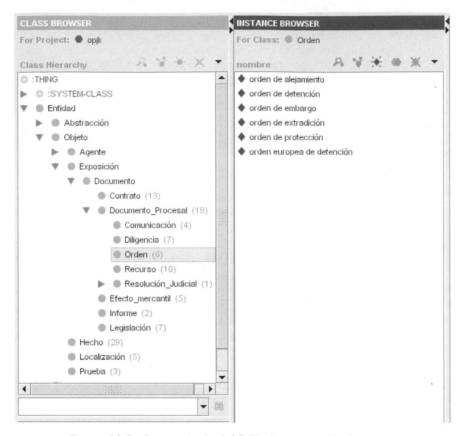

Figure 12.5   Screenshot of OPJK classes and instances.

regarding some terms as significant and their conclusions have proved to be really useful to both feed and control the modeling process (Figure 12.6).

However, linguistic constraints due to the use of the Spanish language within the legal case study, added difficulty to the use of this technology. One of the main problems encountered during the utilization of Text-ToOnto referred to the process of word reduction (i.e., just before the process of concept identification). It uses stemming techniques instead of lemmatization for word reduction, which has proved to be less useful in achieving good results for certain languages.[8]

Stemming works by transforming a word into its stem usually by cutting-off the word suffix. If a stemming process is applied to languages

---

[8] This problem has been widely explained—and a solution proposed—in (Vallbé *et al.*, 2005) and (Vallbé & Martí, 2005).

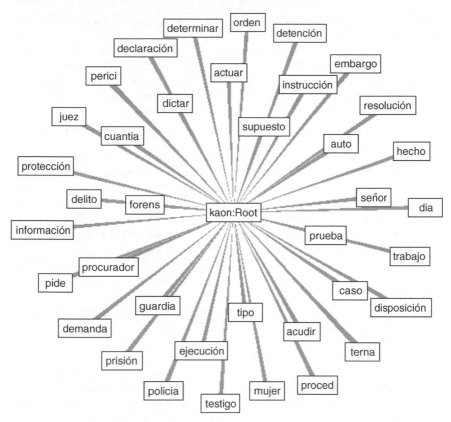

Figure 12.6 Screenshot of the term extraction performed with TextToOnto and visualized with KAON.

such as Spanish, Catalan, or Slovenian, with rich inflection (which can have 60 forms for a verb not counting composed forms) a lot of information keeps hidden and the reduction process based on stemming often produces results that are not refined enough. Moreover, stemming may put multiple forms behind the same stem. Furthermore, in some cases stemming gives different stems when there should be the same stem. This problem has been partially solved by recourse to an open source Spanish lemmatizer,[9] which enables applying a lemmatization process to the corpus before being processed by the tool.

Thus, the main method used in building the ontology focused on the discussion within the UAB legal experts team over the terms that appear on the competency questions. This method had several phases. First, it

---

[9]http://garraf.epsevg.upc.es/freeling/

basically consisted in selecting all the nouns (usually concepts) and adjectives (usually properties) contained in the competency questions.

Once the terms had been identified, the team discussed the need to represent them within the ontology and their organization within taxonomies. The relevant relations between those terms were also identified (mainly is_a and instance_of). Accordingly, we followed the middle-out strategy (Gómez-Pérez *et al.*, 2002). With this strategy, the core of basic terms are identified first and then they are specified and generalized if necessary.[10]

However, difficulties in reaching consensual decisions and the lack of traceable lines of argumentation for both the decisions agreed within the expert's team and the modeling refinement agreed between legal experts and ontology engineers was slowing down the construction of the ontology. For that reason, the introduction of DILIGENT, described in Chapter 9 above, offered a reliable basis for a controlled discussion of the arguments for and against a modeling decision.

12.3.3.2 Construction Methodology

The introduction of DILIGENT not only proved the need to rely on guidelines for the decision-making process within the ontology design, but also facilitated communication between legal experts and ontology engineers in a geographically distributed environment.

The use of DILIGENT sped up the modeling process, as decisions were more easily reached and more concepts were agreed upon. However, the lack of appropriate evaluation measures made it difficult, at times, for the contradicting opinions to achieve an agreement. Although the argumentation stack was captured and tagged after the discussion in order to trace the arguments, an accessible web-based interface was offered in order to track the discussion. A standard wiki was used to support discussion. The ontology discussion wiki made all

---

[10]As an example, and in relation to the competency questions analyzed above, modelers considered that the concepts *auto* [interlocutory decision], *recurso* [appeal], *demanda* [private/civil lawsuit], and *querella* [public/criminal lawsuit] needed to be represented in the ontology. Moreover, a concept *documento* [document] had to be created as all terms: *auto, recurso, demanda,* and *querella* describe documents. The result was the construction of a more general concept from those specific terms found in the competency questions. However, the team also agreed that *demanda, auto, recurso,* and *querella* were not only instances of *documento*, but also constituted a specific class of documents used only within the judicial process. For that reason, *documento_processal*[procedural document] had to be created as a subconcept of *documento*. At the same time, there are different types of appeals and court orders stated in the questions that have to be considered instances of *recurso* and *auto*. In this case, the terms where specified, not generalized. This is a clear example of the use of the middle-out strategy in the legal case study ontology. Furthermore, some other relations (different from is_a and instance_of) were also identified: someone creates those documents (*juez, denunciante, persona*), thus document has_author.

decisions transparent, traceable, and available to all members of the team, especially those joining the team at a later stage.

However, the tool did not provide several features such as: visualization of the graphical representation of the ontology being built or a system of e-mail notifications when arguments had been added. To solve the requirement of graphical visualization, the ontology modeling team extended the wiki with screenshots from the relevant parts of the ontology build with the KAON OI-Modeler.[11] Later, we considered the addition of a referee (or that one of the members of the team played the role of referee) in order to further speed up the discussions and to keep them on track, as discussions often tend to lose focus.

DILIGENT as a methodology facilitated decision-making among the terms and relations that could be included in the ontology.

12.3.3.3 Ontology Integration

Finally, this ontology was integrated into PROTON (ProtoOntology).[12] PROTON is a domain independent ontology and, first, OPJK modelers thought that integration might require some rearrangements, but it was essential for the OPJK to model judicial knowledge as perceived by judges and that point of view has to be maintained when possible.

Finally, OPJK has recently been integrated into the System and Top modules of PROTON (Casellas *et al.*, 2005) and, as top layers represent usually the best level to establish alignment to other ontologies, the classes contained in the Top Module (*Abstract*, *Happening*, and *Object*) were straightforwardly incorporated, together with most of their subclasses, although *Abstract* needed the introduction of a specific subclass *AbstraciónLegal* [LegalAbstraction] for organizational purposes.

Also most of the relations/properties existing between the Top Module classes were inherited. The domain independence of PROTON facilitated the integration of OPJK.

The first part of the integration process consisted mainly in generalizing OPJK concepts taking into account the System and Top modules of PROTON, incorporating the meta-level primitives contained in the System module (i.e., *Entity*) as the application ontology.

Regarding relations, the specificity of the legal (professional) domain requires specific relations between concepts (normally domain-related concepts as well). However, most existing relations between the Top module classes taken from PROTON have been inherited and incorporated. It has not been necessary for the usage of the Iuriservice prototype

---

[11]http://kaon.semanticweb.org/
[12]http://proton.semanticweb.org/

to inherit all PROTON relations, although most of the relations contained in PROTON had already been identified as relations between OPJK concepts.

The following relations—not a comprehensive list—have been inherited from the existing relations in within the Top module concepts: *Entity hasLocation, Happening* has *endTime* and *startTime, Agent* is *involvedIn* (*Happening*), *Group hasMember,* an *Organization* has *parent/childOrganization of* (*Organization*) and is *establishedIn,* and, finally, *Statement* is *statedBy* (*Agent*), *validFrom,* and *validUntil.*

12.4. ARCHITECTURE

12.4.1. Iuriservice Prototype

In this section, we briefly explain the functionalities of the system, provide a high-level overview of the architecture, and provide some initial analysis of the results.

12.4.1.1. Main Functionalities

The system can be best understood as an extended FAQ platform that allows users—judges in our case—to pose a query in natural language, and the systems returns the known questions that best match the user's question. The extension concerns what we call 'answer explanation:' given a particular question-answer pair retrieved from the FAQ repository, users can request supporting documentation for the answer, including judgments and statutes. The key differential aspect of the system is its knowledge about the legal domain. Rather than matching, based on keywords, our system uses ontologies to both retrieve the most similar question and to link to supporting documentation. Figure 12.7 illustrates those two modes; on the left side we see the FAQ part, while on the right hand side the answer explanation functionality is illustrated. As can be seen, the 'answer explanation' part can also be used as a semantic meta-search engine over distributed legal sources.

12.4.1.2. Architecture

In this section, we will provide an overview of the architectures of the two parts of the system.

- *FAQ System*: Several search and score algorithms have been designed based on Natural Language Processing and on Ontology Concepts

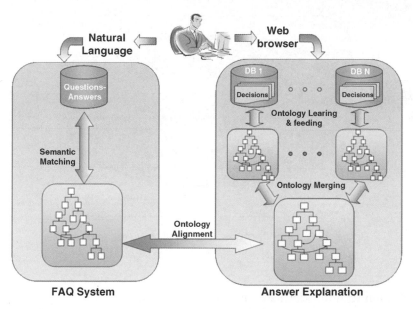

Figure 12.7 High-level architecture of Iuriservice system. A FAQ system is combined with an answer explanation system that provides explanations for the answers provided by the FAQ part.

Matching (Zhu *et al.*, 2002). Algorithms have been organized around an architecture based on an adaptive multistage search chain, which is based on a variation of the 'chain of responsibility' pattern. In particular it is based on a factory pattern that produces, on demand, a suitable search engine. This engine uses some search stage engine plug-ins and adapters to leverage on the main technologies used like NLP processing adapters, Ontology API and algorithms adapters. Each stage behaves independently from previous stages. The stage starts with a FAQ subset as an entry, the goal being to reduce this subset with the constraint that the searched FAQ belongs to it. We have considered a three-stage search process, linking one outcome with the next entry, like a chain of responsibility. The first stage determines the domain of the question such as gender violence, criminal law, etc. The second step uses keyword-based techniques to filter out FAQs that are dealing with other domains than that of the question. In the last stage the semantic distance is determined between the user question and the remaining FAQs. Since this is computationally an expensive process, it will be performed with those stored FAQs whose likelihood of appropriateness is above a certain threshold. Figure 12.8 illustrates this architecture. See (Casanovas *et al.*, 2005a) for details.

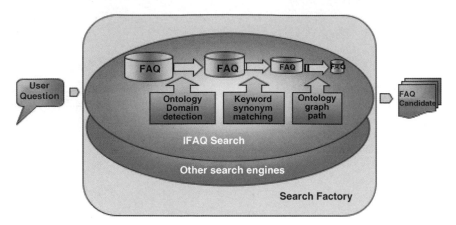

Figure 12.8 Architecture of Iuriservice 'FAQ' subsystem.

The main technologies used in this architecture are:

- Natural Language Processing: NLP is used at several search stages to get additional comprehension from the user's question. A morphological and syntactical analysis of the user's question is performed. The relevant words and grammatical patterns drawn from the question are used by other components in further stages.
- Thesaurus Processing: It is used to match words based on synonymous relationships. The system attempts at both exact and synonym matching.
- Ontology Processing: The system uses several legal domain ontologies to obtain understanding of the user's question. The system tries to find a match between fragments of the user's question and paths in the ontology. To do so, it builds a graph path that is compared to each of the stored FAQ graph paths. We calculate the 'semantic distance' between a new user query and the stored questions. Figure 12.9 illustrates the process of how two ontology fragments are matched to each other.
- Cache Proxy: The system produces intermediate results of repetitive calculations that can be saved to avoid the repetition of computations. Many of these calculations can also be recovered from a repository like a RDBMS and saved on cached memory.
- Answer Explanation System: In the Answer Explanation part of the system, the user can ask for supporting documents for any answer the system offers. In this stage the semantic search engine navigates the case law databases and offers references to relevant documents. This functionality allows the judge to learn from the cases that have originated the answer or precedent. This functionality can also be

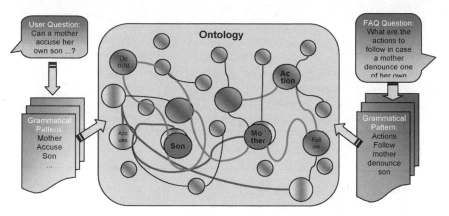

Figure 12.9 Ontology graph matching. Fragments of ontologies are matched onto each other considering semantic distances between concepts.

used directly, as a meta-search engine working upon different case law databases, without the need to ask previously a concrete question. Automatic document processing and understanding based on ontology mapping and alignment technology facilitates the introduction of case law databases into the system repository. In this way, each format representing legal cases is translated into a common schema and processed to establish links to stored FAQ answers. Apart from a standard text interface, we are currently studying intuitive ways to visualize the results. Figure 12.10 shows the Use Case for this subsystem, including a meta-search engine that accesses the case law databases and extracts the relevant information and/or documents, and constructs the explanation. The explanation consists of a set of automatically inserted hyperlinks into the question-answer pairs that point to relevant documents from the case law databases. Databases contain the cases produced by Spanish courts at different levels. (Figure 12.10).

In order to connect the two kinds of knowledge contained in the judicial experience (FAQ system) and the past judicial decisions (Answer Explanation), and to detect the useful cases to justify the answers in the FAQ repository, the concepts in the two main ontologies have to be aligned. Therefore, if a user selects a justification, the system will check the OPJK concepts appearing in the answer, will transform them into the corresponding set of case law ontological concepts, and eventually retrieve the appropriate cases containing those concepts.

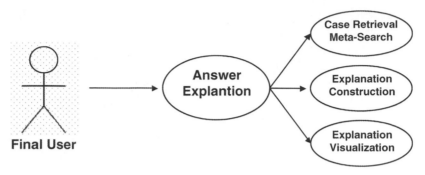

Figure 12.10 Use case for the 'Answer Explanation' subsystem.

12.4.1.2. Initial Results

At the current stage of the project, we have implemented the FAQ subsystem that uses the notion of semantic distance to calculate the similarity between user questions and stored FAQs (question-answer pairs). For this subsystem, we have performed some tests to measure the effectiveness of the retrieval process. The goal of the measurement has been to verify whether retrieval, based on semantic distance technology obtains better results than keyword-based retrieval. To focus the benchmark on the contribution of the 'semantic' part rather than comparing the results to simple keyword–based approaches, we have compared it to an enhanced keyword approach by including a morphological analysis and synonyms. We performed two types of tests using a corpus of 62 FAQs related to the legal area of Gender Violence, whose OPJK ontology contains 82 classes, 118 attributes, and 484 instances.

Same Meaning Test: The first test we performed concerns the retrieval of questions whose intended meaning is similar to that of the user question but different terms are used. The test is a success if the expected user result is the first in the list retrieved by the system. An example of such test is:

User Question: I have ordered an injunction of protection in favor of a woman, and after some days she comes back asking me to cancel or withdraw it. What should I do? Should I withdraw it?

Target FAQ: A woman has come this morning asking for an injunction of protection. We have been busy on this subject for all the day. I have just ordered it, and at this precise moment I am notifying the injunction of protection and she already says that she wants to remove the denounce and that she does not want the injunction. What do I have to do?

In this case, the User Question and the Target FAQ meet the same semantics.

Table 12.1 Summarizes the executed tests.

	Enhanced keywords	Enhanced keywords and semantic distance
Same meaning	35 tests	35 tests
Different meaning	35 tests	35 tests

Different Meaning Test: In this test, we pose a question to the system, knowing that there is a stored question, whose 'keyword characteristics' are similar to the user question, but whose intended semantics are different. The test is a success if the target question is *not* in the list of retrieved questions. An example is:

User Question: Should I order an injunction of protection if a man and a woman live together but the man gets usually very angry and he is also seeing another woman?

Target FAQ: There is a couple and an injunction of protection against the husband, but the police knows that they are living together and they told me that. Any time she gets angry with him or they have some trouble she uses the injunction, the police detains him and I have to organize a hearing··· just to find them together again next morning. What can I do? Can I modify or cancel the injunction?

In this case, the User Question and the Target FAQ have different semantics, and the Target FAQ should not be among the retrieved ones.

We defined 35 test cases for the 'same meaning' type and 35 for the 'different meaning' type. As explained, these tests have been executed by (i) a search engine based on enhanced keyword technology, and (ii) a search engine that additionally takes into account the semantic distance. Table 12.1 summarizes the executed tests.

The results of the tests are summarized in Table 12.2. The table summarizes the percentages of success and failure for the two types of retrieval (keyword vs. semantics) and the two types of tests ('same meaning' vs. 'different meaning'). We can see that the semantic distance

Table 12.2 Summary of test results. Considering the semantic distance improves results in both cases.

		Enhanced keywords	Enhanced keywords and semantic distance
Same meaning	Success	28 57 %	45 71 %
	Failure	71 43 %	54 29 %
Different meaning	Success	17 14 %	40 %
	Failure	82 86 %	60 %

technology gives better results: the number of successes increases while the number of failures decreases.

12.5 CONCLUSIONS

Iuriservice is a prototype of an iFAQ that is being implemented into the judicial Spanish system. A mixed team of Magistrates and researchers have been conducting usability tests and will perform the user validation plan at the Spanish Judicial School with final users in the next future (Bösser, 2005). In this sense, social knowledge and technological knowledge cooperate through the ontological engineering process. An Ontology of Professional Judicial Knowledge (OPJK) and a methodology with a middle-out strategy are being developed. SEKT technologies have been integrated at different stages.

Semantic Web has proved to be very useful in improving the knowledge management skills of the recently appointed judges. Iuriservice is designed not only to be accurate and technologically advanced, but also to fulfil the specific requirements of professional judges:

- It is designed to be efficient, extensible, customizable, and scalable.
- It makes use of incremental search as a process of narrowing the solicited FAQ set.
- It uses a variety of pluggable searching algorithms.
- It is designed to be accurate and technological advanced by using NLP and ontological techniques.

REFERENCES

Ayuso M *et al.* 2003. 'Jueces jóvenes en España, 2002. Análisis estadístico de las encuestas a los jueces en su primer destino (Promociones 48/49 y 50).' Internal Report for the General Council of the Judiciary, within the framework of the Project 'Observatory of Judicial Culture', SEC-2001-2581-C02-01/02.

Benjamins VR, Casanovas P, Breuker J, Gangemi A. 2005. 'Law and the Semantic Web, an Introduction. In *Law and the Semantic Web*, Benjamins *et al.* (ed). Springer Verlag: London, Berlin.

Benjamins VR, Contreras J, Blázquez M, Rodrigo L, Casanovas P, Poblet M. 2004. 'The SEKT Legal use case components: Ontology and architecture'. In: *Legal Knowledge and Information Systems. JURIX 2004: The Seventeenth Annual Conference*, Gordon T F. IOS Press: Amsterdam, pp 69–77.

Blankenburg E. 1999. Legal culture on every conceptual level. In: *Globalization and Legal Cultures*. Feest J. (ed.). IISL Oñati, pp 11–19.

Bösser 2005. Die Analyse der Bedürfnisse und Präferenzen von professionellen Nutzern von Information. 47th meeting of the section Anthropotechnologie of the DGLR - Deutsche Gesellschaft für Luft- und Raumfahrt.

Breuker J, Elhag A, Petkov E, Winkels R. 2002. Ontologies for Legal Information Serving and Knowledge Management. Legal Knowledge and Information

Systems. Jurix 2002: The Fifteenth Annual Conference. Amsterdam, IOS Press, pp 73–82.

Breuker J, Winkels R. 2003. Use and reuse of legal ontologies in knowledge engineering and information management, ICAIL03 *ICAIL 2003 Workshop on Legal Ontologies and Web Based Legal Information Management*, Edinburgh, http://lri.jur.uva.nl/~winkels/legontICAIL2003.html

Casanovas P. 1999. Pragmatics and Legal Culture. ICPS Working Paper n. 159. Barcelona: Institut de Ciències Polítiques i Socials. http://www.diba.es/icps/working_papers/docs/wp_i_159.pdf

Casanovas P, Gorroñogoitia J, Contreras J, Blázquez M, Casellas N, Vallbé JJ, Poblet M, Ramos F, Benjamins VR. 2005a. SEKT Legal Use Case Components: Ontology and Architectural Design. In *Proceedings of ICAIL 05*. ACM, Bologna, 2005, pp 188–194.

Casanovas P, Poblet M, Casellas N, Contreras J, Benjamins VR, Blázquez M. 2005b. Supporting newly-appointed judges: A legal knowledge management case study. *Journal of Knowledge Management* 9(5):7–27.

Casellas N, Blázquez M, Kiryakov A, Casanovas P, Poblet M, Benjamins R. 2005. OPJK into PROTON: Legal Domain Ontology Integration into an Upper-level Ontology. In: *OTM Workshops 2005*, LNCS 3762, Springer-Verlag: Berlin, Heidelberg, Meersman R. *et al.*, (eds). pp 846–855.

Chan CW. 2003. Cognitive modeling and representation of knowledge in ontological engineering. *Brain and Mind* 4:269–282.

Clancey WJ, Sachs P, Sierhus M, Hoof RV. 1998. Brahms: Simulating practice for work systems design. *International Journal of Human-Computer Studies* 49: 831–865.

Eraut M. 1992. Developing the knowledge base: A process perspective on professional education. In: *Learning to Effect*, Barnett R. ed. Open University Press: Buckingham, pp 98–18.

Friedman LM. 1969. Legal culture and social development. *Law and Society Review* 4:29–44.

Friedman LM. 1975. *The Legal System: A Social Science Perspective*. Russell Sage Foundation:

Gangemi A, Breuker J. 2002. *Harmonising Legal Ontologies*. Ontoweb. IST Project 2000-29243. http://ontoweb.aifb.uni-karlsruhe.de/About/Deliverables.

Gangemi A, Pisanelli DM, Steve G. 2001. A formal Ontology Framework to represent Norm Dynamics. *In Proceedings of the Second International Workshop on Legal Ontologies*, Amsterdam.

Gangemi A, Sagri M, Tiscornia D. 2003. Metadata for Content Description in Legal Information, in ICAIL 2003 Workshop on Legal Ontologies & Web based legal information management, June 2003, Edinburgh, Scotland, UK. http://www.lri.jur.uva.nl/~winkels/LegOnt2003/Gangemi.pdf.

Gil R, Grcía R, Delgado J. 2005. *An interoperable framework for Intellectual* Property Rights using web ontologies. In: *LOAIT-Legal Ontologies and Artificial Intelligence Techniques*, Lehmann, J Biasiotti, MA Francesconi, E Sagri MT (eds) Nijmegen: Wolf legal publishers, pp 135–148.

Gómez-Pérez A, Corcho O, Fernández-López, M. 2002. *Ontological Engineering: With Examples from the Areas of Knowledge Management, E-Commerce and Semantic Web (Advanced Information and Knowledge Processing)*. Springer-Verlag: London.

Jarrar M, Meersman R. 2001. Practical Ontologies and their Interpretations in Applications—the DOGMA experiment. http://www.starlab.vwb.ac.be/publications/STAR-2001-04.pdf.

Kralingen van RW. 1995. *Frame-based Conceptual Models of Statute Law*, Computer/Law Series, No. 16. Kluwer Law International. The Hague, The Netherlands.

McCarty LT. 1989. A language for legal discourse, I. Basic features. In *Proceedings of the Second International Conference on Artificial Intelligence and Law*, Vancouver, Canada, pp 180–189.

Menzies T, Clancey WJ. 1998. Editorial: the challenge of situated cognition for symbolic knowledge-based systems. *International Journal of Human-Computer Studies* 49:767–769.

Roche C. 2000. Corporate ontologies and concurrent engineering. *Journal of Material Processing Technology* 107:187–193.

Rodrigo L, Blázquez M, Casanovas P, Poblet M. 2004. D10.1.1. *Before* Analysis. Case Study—Intelligent Integrated Decision Support for Legal Professionals. State of the art. SEKT. EU Project IST-2003-506826.

Stamper RK. 1996. Signs, Information, Norms and Systems. In: *Signs of Work*, Holmqvist, B Andersen P (eds). De Gruyter: Berlin.

Valente A. 1995. *A Modeling Approach to Legal Knowledge Engineering*. IOS Press: Amsterdam, Tokyo.

Valente A. 2005. Types and roles of legal ontologies. In *Law and the Semantic Web. Legal Ontologies, Methodologies, Legal Information Retreval, and Applications*, Benjamins VR *et al.* (eds). LNAI 3369, Springer: Berlin, pp 65–76.

Valente A, Breuker J, Brouwer B. 1999. Legal modeling and automated reasoning with ON-LINE. *International Journal of Human-Computer Studies* 51:1079–1125.

Vallbé J-J, Mortí MA, Fortuna B, Jakulin A, Mladenic D, Casanovas P. 2005. Stemming and lemmatisation Improving knowledge management through language processing techniques. In: *The Regulation of Electronic Social Systems. Law and the Semantic Web*, Casanovas P, Bourcier D, Noriega P, Cáceres E, Galindo F (eds). In *Proceedings of the B4-Workshop on Artificial Intelligence and Law*. IVR' 05-Granada, 25th-27th May. Web location. http://www.lefis.org. XXII World Conference of Philosophy of Law and Social Philosophy. Instituto de Investigaciones Jurídicas, UNAM México [in press].

Visser PRS. 1995. *Knowledge Specification for Multiple Legal Tasks; A Case Study of the Interaction Problem in the Legal Domain*, Computer / Law Series, No. 17, Kluwer Law International: The Hague, The Netherlands.

Visser PRS, Bench Capon, TJM. 1998. A comparaison of four ontologies for the design of legal knowledge systems. *Artificial Intelligence and Law* 6:27–57.

Zhu H. *et al.* 2002. An Approach for semantic search by matching RDF graphs. In *Special Track on Semantic Web at the 15th International Flairs Conference (AAAI)*, May 2002, Florida, USA. http://www.dit.hemut.edu.vn/~tru/SPECIAL-STUDIES/rdf-semantic-matching.pdf

13

A Semantic Service-Oriented Architecture for the Telecommunications Industry

Alistair Duke and Marc Richardson

13.1. INTRODUCTION

Today's telecommunication industry is increasingly competitive with many new entrants to the market and a challenging regulatory environment. Along with the ongoing recovery from the technology boom-and-bust, these factors add up to a tough business environment. Price erosion means that operators (and, in particular, the large incumbents) have realised that they must radically transform the way they do business in order to reduce costs and remain competitive. At the same time, a number of new opportunities and threats are emerging including broadband, WiFi, fixed-mobile convergence, aggressive new market entrants and the blurring of the boundary between IT and traditional telecommunications. Companies are seeking to grow new business, while defending traditional core revenues.

Thus the industry is seeking urgently to reduce IT costs. A Forrester survey (Koetzle, 2001) found that average spending on integration by the top 3500 global companies was $6.3 million and 31% was spent on integrating with external trading partners. There is a focus on faster time to market via more flexible business processes. Furthermore, there is a need to reconfigure system components quickly and efficiently in order to satisfy regulatory requirements for interoperation and to provide fully integrated support systems for increasingly sophisticated services.

Semantic Web Technologies: Trends and Research in Ontology-based Systems
John Davies, Rudi Studer, Paul Warren © 2006 John Wiley & Sons, Ltd

On the other hand, customers are demanding integrated services, tailored to their specific needs. The market is becoming increasingly federated due both to regulatory pressures and to companies' attempts to catch market opportunities with tailored, bundled services. In this market, the number of business to business (B2B) relationships between telecommunications companies and specialist content and service providers has dramatically increased.

All these factors have led many telecommunications companies to radically rethink the way they operate. They have realised that the new environment requires tighter yet more flexible management of processes and the eradication of bureaucracy and duplication of effort and systems. This transformation can be achieved with the adoption of a Service Orientated Architecture (SOA).

The creation of Next Generation Networks (NGN) are being built in BTs 21CN programme will create a single core network capable of carrying many types of network product (e.g., PSTN, Broadband). As well as cost savings, the aim of this network is to be dynamic and flexible in the way it is managed, allowing services to be provisioned much more quickly and easily than before. An important requirement is to have an Operational Support System (OSS) structure that allows this flexibility. Building the OSS based on SOAs will be key to this. The next section, Section 13.2, describes the philosophy behind an SOA. Section 13.3 then explains how Semantic Web Services, described in Chapter 10, can be used to implement an SOA. After that, Section 13.4 describes the role of semantic mediation. Sections 13.5 and 13.6 describe in turn, the use of ontologies in telecommunications standards and the application to a particular case study in the telecommunications industry.

13.2. INTRODUCTION TO SERVICE-ORIENTED ARCHITECTURES

A system based on a Service Orientated Architecture (SOA) is one in which resources are made available to other participants in the network as independent services that are accessed in a standardised way. This provides for more flexible loose coupling of resources than in traditional system architectures (Loosely Coupled, 2005). It permits a move away from point to point integration which is costly and inflexible if carried out on a large scale.

Using an SOA, applications are built around services. A service is an implementation of a well-defined business function, which allows such services to be used by clients in different applications or business processes. Increasingly, organisations are adopting an SOA as a means to enable interoperability and encourage reuse, thereby reducing cost.

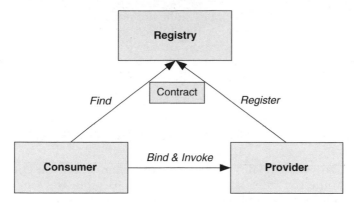

Figure 13.1 The SOA find-bind-execute model.

Services are software components with a well-defined interface that is implementation-independent. A key aspect of an SOA is the separation of the service interface from its implementation (Mahmoud, 2005). The benefits from adopting an SOA approach include:

- Services are self-contained.
- Services are loosely coupled.
- Services can be dynamically discovered.
- Composite services can be built from aggregates of other services.

SOA uses the *find-bind-execute* model as shown in Figure 13.1. Service providers first register their service in a registry. This registry is then used by consumers to find services that match certain criteria. If the registry has such a service, it provides the consumer with a contract and information on accessing the service.

The greater agility afforded by an SOA will also allow organisations to respond to the needs of the market more quickly and in ways that are more attractive to the customer. The SOA is particularly applicable to the Telecommunications market where customer and operational support costs are high and customer satisfaction is a key differentiator.

However, there is evidence to suggest that companies with complex internal organisations and supply chains will find that large scale SOAs are not achievable without semantic descriptions of service components that can aid service discovery and integration. For example, Brodie (2003), Chief Scientist at Verizon Communications stated that:

'There is a growing consensus that Web Services alone will not be sufficient to develop valuable and sophisticated Web processes due the degree of heterogeneity, autonomy, and distribution of the Web. Before the huge promise of Web Services become industry strength, a lot of work is needed, and semantics holds a key'.

It is apparent that Web Services alone are not enough to implement an SOA and enable the advantages that this architecture can bring (such as dynamic discovery and execution of services). Using Semantic Web Services allows the creation of machine readable descriptions of the service capability and interface, allowing the dynamic discovery and execution of services.

13.3. A SEMANTIC SERVICE-ORIENTATED ARCHITECTURE

This section will explain the benefits of semantically described web services in the context of an SOA. In order to do this, the limitations of current web services are first considered.

Web Services are generally described using XML-based standards namely WSDL (which allows one to describe a Web Service in terms of what it does and what its inputs and outputs are), UDDI (which is a centralised registry allowing one to discover Web Services) and SOAP (which is a protocol allowing one to execute services). In addition to these low-level standards, work is in progress to create standards that allow services to be combined into a workflow, for example WS-BPEL (Web Services — Business Process Execution Language) (IBM, 2005) and also to define permissible message exchange patterns and contents, for example ebXML (Eisenberg, 2001). However, none of these standards provide a means to describe a Web Service in terms of explicit semantics. For a given service you might want to describe:

- What kind of service it is;
- What inputs it requires;
- What outputs it provides;
- What needs to be true for the service to execute (pre-conditions);
- What becomes true once the service has executed (post-conditions);
- What effect the service has on the state of the world (and/or the data it consumes and provides).

The first of these requirements is partly addressed by UDDI in that a category and human readable description can be assigned to a web service in a registry to aid discovery. This provides only limited support for automated discovery since a computer will not understand[1] the description or what the category means. The second and third of these requirements are partly addressed by WSDL in that XML tags can be attributed to inputs and outputs. A computer can easily match these but

---

[1]Strictly, the computer never actually understands even when semantics are provided. It is merely provided with the means to relate a piece of information to a machine readable ontology which in turn allows it to determine relationships with other pieces of information and given these perform reasoning to deduce new information. Thus the provision of semantic descriptions makes data much more amenable to machine processing.

again has no notion of their meaning or relationship to other pieces of data. Fundamentally, most of the hard work is left to the human user who must interpret the descriptions provided to the best of his or her abilities.

Services can be described semantically by relating them to ontologies. Ontologies provide a shared view of a domain that can be interpreted by machines. Thus ontologies can describe kinds of services, the data they consume and provide, the processes that services are part of and, equally importantly, the relationships between all of the above.

The explicit relationship between services and ontologies is the key element for Semantic Web Services. It is envisaged that this will enable:

- *Improved service discovery*: Semantic Web search technology allows users to search on ontological concepts rather than by keywords. A simple keyword search only finds where a particular term occurs, and does not give details about its context or relationship to other information. Ontological searches utilise the structured way that information is modelled to allow more powerful searches, such as the ability to query attributes or relationships between concepts. This will allow users (and indeed computers) to find the most appropriate services more quickly or narrow down their search via more expressive queries if required.
- *Re-use of service interfaces in different products/settings*: Services that are described semantically can more easily be discovered, understood and applied thus reducing the need to create new services that serve the same purpose. This could also be used in a strategy to reduce complexity, that is remove services/interfaces that exactly repeat the function of other services but are described slightly differently.
- *Simpler change management*: Changes to models and services are inevitable over time. The key thing is to reduce the knock-on effect of change or at least manage it. A semantic approach will significantly reduce the overhead and simplify the process. For example, when a proposed change is made to a data element, those services or interfaces that employ that data in some way can be dynamically discovered and appropriate action could be taken, for example to contact the owner of the service with details of the proposed change.
- *A browseable, searchable knowledge base for developers (and others)*: In tandem with the example given above for simpler change management, semantically described services and ontologies enable a knowledge base to be constructed. This allows developers and solution providers to perform queries relating to the data and processes they are concerned with, for example to determine the origin or destination of a piece of data.
- *Semi-automatic service composition*: Given a high level goal which we wish a service or set of services to achieve, expressed in terms of an ontology, it is possible to carry out decomposition into component parts and then match these components with appropriate services. The

level of automation possible is a matter for ongoing research. Initial practical results are likely to provide users with a set of candidate services that might satisfy their needs. They are then left to decide between these services and oversee the composition required in order to satisfy the goal.

- *Mediation between the data and process requirements of component services*: Often there is need for two or more services to interact even though their communication requirements are semantically the same but syntactically different (they may require different message exchange patterns or different data formats). In this case it should be possible to automatically construct a translation between message data elements that allows the services to communicate. This is an example of a process known as mediation, which is discussed in more detail in the next section. It relies upon the mappings of messages and data elements to an ontology allowing semantic equivalence to be inferred.

- *Enterprise Information Integration*: As the name suggests, the Semantic Web builds upon existing Web technology. This can afford universal (or at least enterprise-wide) access to semantic descriptions of services (or information). One advantage is the ability to construct complex queries which can be executed over a variety of heterogeneous systems. For example, suppose there is a requirement to determine the number of customers within a particular postcode who spend more than £100 per quarter. If that information is held within one database and the person asking has access to it and knows how to query it then an answer could readily be obtained. Of course the situation is more complex if multiple databases hold the answer and access and a query interface have to be determined. The humans involved have some work to do in locating the data and processing it in the required way. A semantic approach, however, allows a single query to be made via a unifying ontology.

13.4. SEMANTIC MEDIATION

The role of mediation in supporting an SOA has already been noted. Mediation is generally achieved through the use of mediators, that is components which enable heterogeneous systems to interact. In a practical sense, mediators have generally been realised as pieces of program code that perform point-to-point, low-level translations. Although such mediators satisfy the short-term goal in that they allow two systems to talk to each other, they suffer from maintainability and scalability problems. In general, it is not likely to be feasible to automate their application in a dynamic environment because of their close coupling with the implementation.

Semantic Mediation enables a more dynamic approach through the use of ontologies, which provide consensual and formal conceptualisation of a given domain. 'Mediators can be used to convert from a source implementation interface to that of a target implementation. Modelling the processes and data in the source and target interfaces using ontologies, enables the definition of relationships between semantically equivalent concepts. The mediator can use these relationships to dynamically map between the source and target'.

Mediation can be classified as acting on both data and process. The following two sections describe this in more detail.

13.4.1. Data Mediation

Data mediation is required when the semantic content of a piece of data or message provided by one system and required by another is the same, but their syntactic representations are different. This may be due to differing naming or formatting conventions employed by the partner systems. In order to overcome these mismatches, a mapping tool can be applied at design time. These can be used to map source elements to target elements, often on a one-to-one basis. Where more complex mappings are required such as many-to-one mappings or mappings that are dependent upon content, a rule language may be necessary to describe them. Once a data mediator has been developed its functionality should be described (e.g. the source and target that it mediates between) so that interested parties (be they humans or computers) can inspect it and use if necessary.

13.4.2. Process Mediation

Process mediation is required when the semantic content of a process is shared by two parties but the messages or message exchange patterns of the parties required to achieve that process differ. The process mediator must ensure that the message exchange required by each party is adhered to. As a result the mediator may need to, for example, create new messages that appear to come from the source party and send these to the target. The content of such created messages would have been obtained from the source by the mediator either by explicitly asking for it or by retaining it until required by the target.

13.5. STANDARDS AND ONTOLOGIES IN TELECOMMUNICATIONS

The Telecommunications Industry is seeking ways to encourage interoperability among the many systems required to run and manage a telecommunications network. One such approach is the New Generation Operations Systems and Software (NGOSS) initiative from the TeleManagement Forum (TeleManagement Forum, 2005a). NGOSS is an integrated framework of industry agreed specifications and guidelines which include a shared information and data model for systems analysis and design, and a process framework for business process analysis. NGOSS is intended to allow easier integration of the Operational Support Systems (OSS) software used to provision, bill and manage network-based products and services.

Part of the work of NGOSS is to produce standards for Next Generation Networks (NGNs). Currently telecommunications companies have many different networks for different services (e.g. PSTN, Leased Line) that require managing and maintaining individually. This requires hundreds or even thousands of different bespoke system for each network to enable billing, maintenance, trouble reporting etc. Telco's are moving towards a consolidated IP-based core to their networks, where many network services can be provided over one core network. This should lead to substantial cost savings and greatly improve flexibility and efficiency in providing network services.

NGOSS has identified that the use of SOA will be important in managing the NGNs as the benefits offered by SOAs fit well into the dynamic and highly flexible architecture that NGNs offer. The critical features of an SOA are captured in the NGOSS principles:

- *Shared Information Data Model*: NGOSS components implement and use a defined part of the Shared Information/Data Model (SID) (Telemanagement Forum, 2005b).
- *Common Communications Vehicle*: Reliable distributed communications infrastructure, for example software bus integrating NGOSS components and workflow.
- *External Process Control*: Separation of End-to-End Business Process Workflow from NGOSS Component functionality.
- *Business Aware NGOSS Components*: Component services/functionality are defined by NGOSS Contracts.

The work of the TeleManagement Forum in developing a framework for Next Generation OSS can be seen as ontology building in that NGOSS provides a level of shared understanding for a particular domain of interest. NGOSS (TeleManagement Forum, 2005a) is available as a toolkit of industry-agreed specifications and guidelines that cover key business and technical areas including Business Process Automation and Systems Analysis and Design. The former is delivered in the enhanced Telecom Operations Map (eTOM^{TM}) (TeleManagement Forum, 2005c) and the latter is delivered in the SID. The eTOM provides a framework that allows processes to be assigned to it. It describes all the enterprise processes required by a service provider. The SID provides a common vocabulary allowing these processes to communicate. It identifies the entities involved in OSS and the relationships between them. The SID can therefore be used to identify and describe the data that is consumed and produced by the processes.

13.5.1. eTOM

The eTOM can be regarded as a Business Process Framework, since its aim is to categorise the business activities embodied in process elements so that these elements can then be combined in many different ways, to implement end-to-end business processes (e.g., billing) which deliver value for the customer and the service provider.

The eTOM can be decomposed to lower level process categories, for example 'Customer Relationship Management' is decomposed into a number of categories, one of which is 'Problem Handling'. This is then decomposed further into categories such as 'Track and Manage Problem'. It is to these lower level categories that business specific processes can be mapped. eTOM uses hierarchical decomposition to structure the business processes. Process elements are formalised by means of a name, a description, inputs/outputs and a set of known process linkages (i.e., links to other relevant categories).

The eTOM supports two different perspectives on the grouping of the detailed process elements:

- Horizontal process groupings, in which process elements describe functionality that spans horizontally across an enterprise's internal organisations (e.g., market, product, customer and service management etc.).
- Vertical process groupings, in which process elements are grouped within End-To-End processes (e.g., fulfilment, assurance etc.) accomplished by the Service Provider enterprise.

The eTOM Business Process Framework is defined as generically as possible, so that it is independent of organization, technology and service.

13.5.2. SID

The SID is much more complex than the eTOM in both its aims and form. It provides a data model for a number of domains described by a collection of concepts known as Aggregate Business Entities. These use the eTOM as a focus to determine the appropriate information to be modelled. The SID models entities and the relationships between them. For example a 'customer' is defined as a subclass of 'role'. It contains attributes such as 'id' and 'name'. It is linked to other entities such as 'CustomerAccount' with an association 'customerPossesses'.

13.5.3. Adding Semantics

Although the TMF NGOSS is one of the more prominent initiatives in standardising data and process models for telecommunications, there are also other attempts from different groups in the industry such as ITU-T (2005), 3GPP (2005) and IPNM (2005). It is Important for NGN to be based on standardised data models but it is unlikely that one particular model will be mature enough to implement in the next 2–3 years (the timeframe for deploying the first generation of NGN).

Ontologies provide a solution due their flexibility in modelling and the ability to easily mediate between ontologies representing different data models. This allows a single conceptual view over several data models. In the classical approach, data models represented in a format such as XML would not easily allow mappings to be defined between them, or allow remodelling and adjustment as the standards develop over time.

For the first step in adding semantics to the NGOSS it was decided to concentrate only on the SID and eTOM as these most closely fit the requirements for building a Semantic SOA prototype based around common OSS assurance tasks. Given that ontologies are a conceptualisation of a domain and the Web Services Modelling Ontology (WSMO, 2005) is a specific form of ontology intended to represent services, their capabilities and data requirements; it is natural to represent the SID and eTOM in WSMO as domain ontologies for data and process. Ontologies are the key element of WSMO since the other three elements (Web Services, goals and mediators) all refer to them. Representing SID and eTOM ontologically will enable service components in the SOA to be described as Web Services using WSMO, with descriptions that refer to the domain ontologies. Similarly WSMO goals for web service discovery can be expressed in the same terms. Mediators will make use of the domain ontologies to, for example, enable mappings between the different message formats of two communicating services. The use of WSMO in this context creates an explicit link between a capability described in a model and the actual service component that will provide it. Subsection 13.6.3.1 gives more information on how the SID and eTOM were used as domain ontologies in the case study prototype.

13.6. CASE STUDY

Although the first application of SOAs has generally been within the boundaries of companies, the benefits equally apply where it is required to integrate the services of customers, suppliers, partners etc. The longer-term vision is that Web Services will compete and collaborate over the Internet and that businesses will trade with partners and with consumers based upon highly dynamic commercial arrangements (Muschamp, 2004). Prior to this vision being realised, SOAs can already be used where trading partner agreements already exist and this is the focus of our case study.

Traditionally, vertically integrated telecommunications companies such as BT have provided end-to-end services to customers using their own retail operations and their own hardware. Over recent years, these companies have worked hard to improve customer service and reduce costs through greater process efficiency and effectiveness. These efforts have been enhanced with the introduction of integrated Operational Support Systems (OSS). These can provide customers with end-to-end visibility of service delivery and assurance. The challenge in the new environment is to maintain these levels of efficiency and customer service even though the service is being delivered by multiple parties and organisations who inevitably have their own systems that cannot be directly integrated with those of others (Evans, 2002). BT Wholesale's B2B Gateway is provided to Service Providers[2] to allow them to integrate their OSS with those of BT. Without such a system the service provider would either need to manually coordinate with BT via a BT contact centre or operate a system separate to its own OSS that communicated with BT's—thus requiring information to be entered twice.

The B2B Gateway exposes an interface which is a combination of transport technologies such as SOAP, security protocols such as SSL, and messaging middleware such as ebXML, and linked to the behaviour of back-end systems. Messages formats are expressed using XML Schema (XSD) (The World Wide Web Consortium, 2000) which has the advantage of availability of tools and the increased possibility of integrating with newer transport standards such as Web Services.

Currently the process involved in granting access for a new service provider on the Gateway is lengthy and complex. It commences with a communication phase where partners assess their technical suitability, receive documentation and consider the level of fit with their existing OSS. A development phase follows, during which support is provided by BT. During the testing phase, the partner is given access to a test environment provided by BT where they can test the validity of their messages and their transport and security mechanisms. Firewalls, proxies etc. must be configured by both parties to ensure that communication can occur. Once the testing phase is complete and documented the partner can move to a pilot phase where terms must first be agreed regarding volumes, frequency and support arrangements before access is

---

[2]A service provider in this context is the organisation which has the relationship with the end customer.

given to the live system. Transactions are monitored during the pilot phase to ensure validity.

The Gateway currently exposes a number of interfaces concerned with service fulfilment and assurance. These are generally concerned with regulated services such as broadband access. The interfaces allow Service Providers to order and cease broadband lines on behalf of their customers, manage faults (i.e. raise faults, request, confirm and cancel repair appointments and receive fault status notifications) and carry out diagnostics (i.e., request tests and handle the response to these).

The process can take several months from start to finish. Any approach that can reduce development time, improve the quality of development through enhanced understanding, and as a result avoid significant problems during the testing and pilot phases will naturally save BT and its partners significant time and money. The remainder of this section will examine how, by using Semantic Web Services, these goals can be achieved for one particular function, that of Broadband Diagnostics.

13.6.1. Broadband Diagnostics

As part of its OSS process, a Service Provider may wish to raise a test on the BT network. This is typically due to a problem that has been reported by one of its customers. The Service Provider's OSS should collect the necessary information from the customer and, assuming that the problem cannot be resolved internally, issue a request via the B2B Gateway.

Interactions are implemented through the exchange of business documents, sent as messages. These interactions are known as transactions. The Gateway currently uses ebXML Business Process Specification Schema (ebXML, 2003) to model the sequencing of these transactions in a collaboration. The Broadband Diagnostics interface has only two transactions. These are 'RequestTest' and 'NotifyOfTestCompleted'. 'RequestTest' is a 'RequestResponse' transaction which means that a response to the test request is expected. This response indicates whether the test has been accepted or rejected. It may be rejected if, for example, the Service Provider is requesting a test on a circuit which it does not own. The 'NotifyOfTestCompleted' is a 'Notification' transaction. This is a single message that is sent following the completion of an accepted test describing the results of the test.

13.6.2. The B2B Gateway Architecture

The B2B Gateway, in common with most B2B interfaces has three separate elements. The two internal systems of the respective organisations that need to communicate and the interface that they will use to do

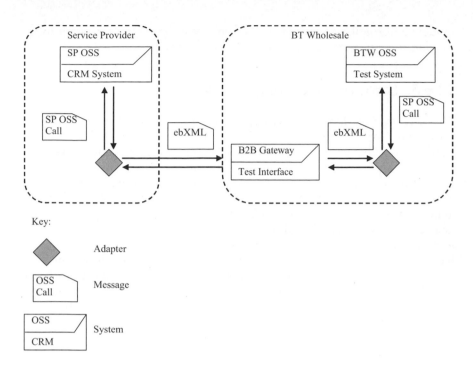

Figure 13.2 B2B gateway architecture.

this. This usually involves both systems translating their internal application view of data and process into the interface view of the problem. Depending upon who produces the interface definition, the amount of translation involved can be either very small or almost impossible to achieve without development effort.

The Gateway architecture can be represented as shown in Figure 13.2. The Service Provider's OSS is able to generate a call to request a test. In order to pass this on to the B2B Gateway, it must first be adapted to enable it to be understood. The adaptation process has two key elements. First, the test call must be represented as a business message that will be understood by the gateway as valid, given the current state of the transaction. That is, it must be represented as a TestRequest message which is the initial interaction of the 'RequestTest' transaction. Second, the business message must be wrapped within the protocol envelope, that is ebXML messaging. A message received by the B2B Gateway must also be adapted before it can be processed by the BT Wholesale OSS. This adaptation is effectively the reverse of the previous one.

Generating the adapter between OSS calls and valid B2B Gateway messages is one of the key challenges of the integration process. The Web Services Modelling Ontology aims to significantly simplify this integration process. The next section describes a prototype using WSMO to model the broadband interface, allowing ontological representations of the data being exchanged to enable semantic mediation.

13.6.3. Semantic B2B Integration Prototype

This section describes the prototype system—The B2B Integration Platform—developed to allow mediation to occur between the Service Provider trading partner and the B2B Gateway. The prototype is based upon the execution environment of the Web Services Modelling Ontology—WSMX (WSMX, 2005). The components of this architecture include Process Mediation (the task of resolving heterogeneity problems in communicating processes) and Choreography (the task of semantically describing the expected message-exchange patterns), which is required by process mediation. Adaptor components have been added to allow low level messages to be represented in WSML (Web Services Modelling Language), the language associated with WSMO and which can be interpreted by WSMX. In this specific use case, multiple Service Providers are interfacing with one Wholesale Provider (BT).

13.6.3.1. Design-Time

The prototype relies upon a number of design-time activities that must be carried out in order for mediation to occur at run-time. From BT's point of view, the key design-time task is to represent its interfaces semantically. This includes adapting the message descriptions to the language of the platform—WSML. It is envisaged that a library of adaptors will exist to convert to and from popular messaging formats such as ebXML, UBL [Oasis] etc. No intelligence is required in this adaptation step and the result is an ad hoc messaging ontology that models the elements of the messages in WSML. Following the adaptation, the elements can then be referenced against a domain ontology, in this case using the industry standard specification Shared Information/Data Model of the TeleManagement Forum (TeleManagement Forum, 2005c). These references provide context to the data and allow their semantic meaning to be inferred. For example the SID defines two concepts `Party` and `PartyRole`. The concept Party is used to explicitly define an organisation or individual and `PartyRole` allows an organisation/individual to take on a particular role during a business transaction. On the B2B Gateway these concepts fit nicely, as there are a number of organisations that use the Gateway (such as BT and other third party providers) and take on different roles depending on the operation being undertaken. If a third party provider wishes to carry out a `testRequest` operation, then the Concept `Party` is used to describe their organisation, and `PartyRole` is used to define their role in this transaction as 'Conductor'. Similarly BTs

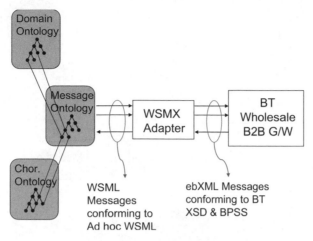

Figure 13.3 BT design-time tasks.

partyRole in this operation is 'Perfomer' as they are performing the actual test.

The final design-time task for BT is to semantically describe the message-exchange pattern that it expects. As explained previously, this is known as choreography. The choreography relates the semantic content of the messages to a semantic description of the process. This can be used by a process mediator to reason about how to mediate to a target choreography. The design-time tasks for BT are illustrated in Figure 13.3.

From the perspective of the Trading Partner, the design-time activities include applying an appropriate adaptor to their message descriptions, defining its own semantic choreography description and defining a data mediator between its data representation and that of BTs. This final step is perhaps the most important and labour intensive. However the open architecture should allow discovery and reuse of mediators if they already exist. The end result of this mediation step is that the *ad hoc* messaging ontology of the Trading Partner is mapped to the domain ontology enabling semantic equivalence. A data mediator is produced that is stored and applied at run-time. The mediator acts as a declarative transform that can be dynamically discovered and applied in other (perhaps closely related) scenarios. As such, it should be stored in such a way that other parties can later discover it.

The choreography of the Trading Partner can be compared with the choreography of BT by the Process Mediation system which can reason whether it is possible to mediate and if so, automatically generate a process mediator. This reasoning step can be carried out at design-time if the two parties are known at this stage (as is the case here) or at run-time if one of the parties discovers the other in a dynamic run-time scenario as described in Section 13.1. This latter case is only feasible if data mediation

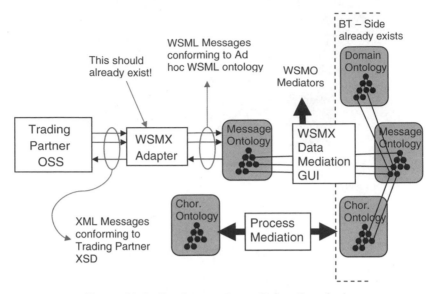

Figure 13.4 Trading partner design-time tasks.

has already occurred or a suitable data mediator can be discovered (Figure 13.4).

13.6.3.2. Run-Time

The sequence of events at runtime are:

1. The trading partner OSS generates a message in its native format, for example XML and forwards this to the Integration Platform.
2. The Integration Platform applies the appropriate adaptor to convert the message to WSML.
3. A description of the appropriate target interface is retrieved from the data store of the platform. This can either be predetermined at design-time or discovered at run-time in a more flexible scenario.
4. The choreography engine identifies suitable process and data mediators for the message exchange.
5. If it is appropriate to send an outgoing message to the target system at this stage, the choreography engine applies the data mediator to generate a message that the target will understand.
6. The outgoing message is adapted to the native format of the target interface. In this case, the target interface is that of the B2B Platform, which is ebXML.
7. The outgoing message is forwarded to the intended destination.

Of this sequence, steps 2–6 are platform-dependent in that they are carried out by the WSMX architecture. However, it is worth pointing out that the key benefit is obtained by the explicit relation that is made between the low-level messages and the domain ontology. Any platform able to interpret this relationship would be able to apply mediation, thereby transforming the data and process to that required by the target.

13.6.4. Prototype Implementation

The prototype has been implemented using WSMX components to form the B2B Integration platform. Web-based GUIs, backed by appropriate Web Services, simulate the OSS of the ISP and BT Wholesale. The web services observe the behaviour of the working systems in that actual message formats and exchange patterns have been utilised. The following describes the RequestTest process that has been implemented for the Assurance Integration scenario. A screenshot from a trading partner GUI is shown in Figure 13.5.

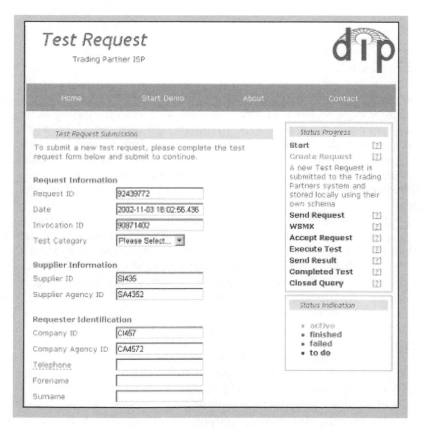

Figure 13.5 Screenshot from prototype UI.

1. A customer informs his ISP of an error occurring in one of his products through a form on the ISP's web site. The error is passed to the ISP's trouble ticketing system.
2. The ticketing system raises the problem with an operator who uses the GUI of the OSS (as shown in Figure 13.5) to request that a test should be carried out on the customer's line. The OSS system produces a message in a specific XML format (including the data payload, describing the error and the customer's product).
3. The message is sent to the B2B Integration Platform which carries out the steps described in Subsection 16.3.6.2 resulting in a test request being forward to BT.
4. BT's OSS receives the message and handles it appropriately, updating its GUI with details and status of the test.
5. Upon completion of the test, the status is updated and an appropriate message is returned to the B2B Integration Platform which again carries out the steps described in Subsection 16.3.6.2. This results in a test request response being sent to the ISP which then updates its GUI allowing the operator to see the result and act on it.

13.7. CONCLUSION

The prototype described is a first step in demonstrating how the goals of an SOA can be assisted with the use Semantic Web technologies.

The main aim of SOAs is to encourage the reuse of available services and allow the flexibility to quickly build complete systems dynamically from the available resources. This has been partially demonstrated in the prototype by showing how ontologies and Semantic Web Services can provide a dynamic and flexible way of integrating services.

Looking ahead, many more players within the industry are expected to expose their interfaces for integration. These will include service, wholesale and content providers. In this scenario, dynamic integration technologies such as WSMO have real value since the economies of scale are greater. The initial effort required in creating ontologies, describing interfaces semantically and relating the two together is much less than the total integration effort. It is also likely that certain ontologies will flourish while others will not, resulting in *de facto* standard ways of describing things. Mediation will be important both to map low level messages and data to the ontologies; and also because new services will emerge requiring integration between the services (and ontologies) of players in previously unimagined fields.

A further aim is to show how semantic descriptions can enable services to be dynamically discovered, composed and executed at runtime. This will be demonstrated in a second prototype.

REFERENCES

3GPP. 2005. *The 3rd Generation Partnership Project* [Online]. Available on the web at: http://www.3gpp.org/

Brodie M. 2003. *The Long and Winding Road To Industrial Strength Semantic Web Services* [Online]. Keynote Talk. ISWC 2003. Available on the web at: http://iswc2003. semanticweb.org/brodie.pdf

ebXML. 2003. *The Definition of Business Processes* (2003) [Online]. Available on the web at: http://www.ebxml.eu.org/process.htm

Eisenberg B, Nickull D. 2001. *ebXML Technical Architecture Specification v1.04* [Online]. Available on the web at: http://www.ebxml.org/specs/ebTA.pdf

Evans D, Milham D, O'Sullivan E, Roberts M. 2002. Electronic gateways—forging the links in communications services value chains. *The Journal of The Communications Network.* 1(1).

IBM. 2005. *Business Process Execution Language for Web Services version 1.1* [Online]. Available on the web at: http://www-106.ibm.com/developerworks/webservices/library/ws-bpel/

IPNM. 2005. *The IP Network Management project* (2005) [Online]. Available on the web at: http://www.tmforum.org/browse.asp?catID=2013

ITU. 2005. *Telecommunication Standardization Sector* [Online]. Available on the web at: http://www.itu.int/ITU-T/

Koetzle L, Rutstein C, Liddell H, Buss C. 2001. *Reducing Integration's Cost.* Forrester Research, Inc.

Loosely Coupled Website. 2005. *Glossary Definition of SOA.* [Online]. Available on the web at: http://looselycoupled.com/glossary/SOA

Mahmoud Q. 2005. *Service-Oriented Architecture (SOA) and Web Services: The Road to Enterprise Application Integration (EAI)* [Online]. Available on the web at: http://java.sun.com/developer/technicalArticles/WebServices/soa/

Muschamp P. 2004. An introduction to Web Services. *BT Technology Journal* 22.

Oasis. *OASIS Universal Business Language (UBL)* [Online]. Available on the web at: http://www.oasis-open.org/committees/tc home.php?wg_abbrev=ubl

TeleManagement Forum. 2005a. *NGOSS Overview Document* [Online]. Available on the web at: http://www.tmforum.org/

TeleManagement Forum. 2005b. *Shared Information/Data Model (SID)* [Online]. Available on the web at: http://www.tmforum.org/

TeleManagement Forum. 2005c. *Enhanced Telecom Operations Map (eTOM) data sheet* [Online]. Available on the web at: http://www.tmforum.org/

The World Wide Web Consortium. 2000. *XML Schema* [Online]. Available on the web at: http://www.w3.org/XML/Schema

WSMO. 2005. Web Service Modeling Ontology (2005) [Online]. Available on the web at: http://www.wsmo.org/TR/d2/v1.2/

WSMX. 2005. *Web Service Modelling eXecution environment* (2005) [Online]. Available on the web http://www.wsmx.org

14

Conclusion and Outlook

John Davies, Rudi Studer, Paul Warren

The chapters of this book provide a comprehensive overview of the current state of the art of ontology-based methods, tools, and applications. They clearly indicate that the progress made in developing Semantic Web methods have resulted in technologies that are applicable in real-world scenarios and provide obvious added value to the end users when compared to traditional solutions.

However, when investigated in some technical detail, one can easily see that the development of semantic applications is largely based on a single or very few related ontologies which are used in a 'one-size-fits-all' approach. Aspects of contexts (such as, e.g., user preferences) that require the use of related yet partially inconsistent ontologies, aspects of networked ontologies dynamically adapting to their changing environment or to the evolving user needs, or aspects of tailoring the human-ontology interaction to specific tasks and users' profiles have not yet been addressed satisfactorily. Furthermore, the semantic handling of resources is more or less constrained to textual resources and, thus, the semantic analysis of multimedia resources is still a challenging issue. These issues are closely related to the fast growing demand of knowledge workers for better management of their personal information on their respective desktops.

Below, we address these open issues in more detail.

14.1. MANAGEMENT OF NETWORKED ONTOLOGIES

Next generation semantic applications will be characterized by a large number of networked ontologies, some of them constantly evolving,

Semantic Web Technologies: Trends and Research in Ontology-based Systems
John Davies, Rudi Studer, Paul Warren © 2006 John Wiley & Sons, Ltd

most of them being locally, but not globally, consistent. In such scenarios it is more or less infeasible to adopt current ontology management models, where the expectation is to have a single, globally consistent ontology which serves the application needs of developers and possibly integrates a number of pre-existing ontologies.

What is needed is a clear analysis of the complex relationships between ontologies in such networks, resulting in a formal model of networked ontologies that supports their evolution and provides the basis for guaranteeing their (partial) consistency in case one of the networked ontologies is changing. Open issues that are involved are among others:

- Notion of consistency: The notion of consistency which is appropriate in this network of ontologies in order to meet the requirements of future real-life applications needs to be analyzed.
- Evolution of ontologies and metadata: One has to investigate which kind of methods are suitable for supporting the evolution of these networked ontologies. Here, one has to analyze the impact of centralized versus decentralized control mechanisms, especially when scalability has to be taken into account. Furthermore, one has to coordinate the evolution of networked ontologies with the evolution of the related metadata. Since networked ontologies will result in collections of metadata that are distributed as well, the synchronization of evolution processes in these distributed environments requires the development of new methods that are able to cope with these distribution aspects.
- Reasoning: A basic open issue is the development of reasoning mechanisms in the presence of (partial) inconsistencies between these networked ontologies. Whereas first solutions have been developed that provide basic functionalities, the main challenge is still how to come up with methods and tools that scale up to handle a large number of networked ontologies and related metadata.

Developing methods and tools that are able to meet these challenges is an essential requirement to devise an ontology and metadata infrastructure that is powerful enough to support the realization of applications that are characterized by an open, decentralized, and ever changing environment.

14.2. ENGINEERING OF NETWORKED ONTOLOGIES

In recent years several methodologies have been developed to engineer ontologies in a systematic and application driven way. However, when considering the needs of ontology engineers and ontology users various aspects of ontology engineering still need significant improvement:

- Semi-automatic methods: The effort needed for engineering ontologies is up to now a major obstacle to developing ontology-based applications in commercial settings. Therefore, the tight coupling of manual methods with automatic methods is needed. Especially, the integration of methods from the area of information extraction on the one hand and from the area of machine learning on the other hand still needs improvement. Here, a deeper understanding of the interplay of these methods with the semantic structures as provided by ontologies is needed. In essence, such an understanding would provide guidelines for a more fine-grained guidance on how to use these automatic methods depending, for example, on the nature of resources available or the usage behaviour of the application users.

- Design patterns: Analogous to the development of design patterns in software engineering, the engineering of ontologies has to be improved by the development of pattern libraries that provide ontology engineers with well engineered and application proven ontology patterns that might be used as building blocks. Whereas initial proposals for such patterns exist, a more systematic evaluation of ontology structures and engineering experiences is required to come up with a well-defined library that meets the needs of the ontology builders.

- Design rationales and provenance: With respect to maintaining and reusing ontologies, methodologies have to provide a more comprehensive notion of design rationales and provenance. When thinking of networked scenarios where ontologies are reused in settings that had not been envisioned by the initial ontology developers, providing such kinds of metainformation about the respective ontology is a must. Here, there is a tight dependency with regard to the above-mentioned use of automatic methods, since, for example, provenance information has to be provided along with the generated ontology and metadata elements.

- Economic aspects: In commercial settings, one needs well-grounded estimations for the effort one has to invest for building up the required ontologies in order to be able to analyse and justify that investment. Up to now, only very preliminary methods exist to cope with these economic aspects, typically constrained to centralized scenarios. Since good estimations depend on many parameters that have to be set for a concrete application scenario, improvement in this area also heavily depends on collecting experience in real-life projects, comparable to the experience that is the basis for these kind of estimations in the software engineering area.

Thus, although the engineering of ontologies is a research area already receiving considerable attention, there still exist a significant amount of open issues that have to be solved for really meeting the needs of developers of ontology-based applications.

14.3. CONTEXTUALIZING ONTOLOGIES

Since ontologies encode a view of a given domain that is common to a set of individuals or groups in certain settings for specific purposes, the mechanisms to tailor ontologies to the need of a particular user in his working context are required. The efficient dealing with a user's context posts several research challenges:

- Formal representation of context: Context representation formalisms for ontologies should be compliant with most of the current approaches of contextual modeling from more traditional logical formalisms to modern probabilistic representations. Such formalisms should also support descriptions of temporal contexts in order to deal with context evolution.
- Context reasoning: Reasoning processes can be used to, among other things, infer the same conclusions from different ontologies using different contexts, to draw different conclusions from the same ontologies using different contexts, or to adapt an ontology with regard to a context and to deal with such a modified ontology. Practical reasoning with contexts should encompass methods for reasoning with logical representations (such as description logic) on one side and probabilistic representations (such as Bayesian networks) on the other side of the spectrum. Special attention should be given to the scalability of the approaches.
- Context mapping: Interoperability between different contexts in which an ontology is used can be achieved by the specification of mappings that formalize the relationships between contexts. The formal specification of such context mappings might support the automatic analysis of these context dependencies, like, for example, consistency. Using terminological correlations, term coreferences, and other linguistic and data analysis methods it might be possible to at least partially automate the creation of mappings between contexts, thus decreasing the required human involvement in the creation and use of contextualized ontologies.

A promising application area of contextual information is user profiling and personalization. Furthermore, with the use of mobile devices and current research on ubiquitous computing, the topic of context awareness is a major issue for future IT applications. Intelligent solutions are needed to exploit context information, for example, to cope with the fuzziness of context information and rapidly changing environments and unsteady information sources. Advanced methodologies for assigning a context to a situation have to be developed, which pave the way to introduce ontology-based mechanisms into context-aware applications.

14.4. CROSS MEDIA RESOURCES

More and more application scenarios depend on the integration of information from various kinds of resources that come in different formats and are characterized by different formalization levels. In a lot of large companies, for example, in the engineering domain, information can be typically found in text documents, e-mails, graphical engineering documents, images, videos, sensor data, and so on, that is, information is stored in so-called cross-media resources. Taking this situation into account, the next generation of semantic applications have to address various challenges in order to come up with appropriate solutions:

- Ontology learning and metadata generation: Methods for the generation of metadata as well as the learning of ontologies have until now been focused on the analysis of text documents, information extraction from text being the area of concern. However, since these other kinds of resources are increasingly prevalent, methods, and tools are urgently needed for the (semi-)automatic generation of metadata or the learning of ontologies from these nontextual resources. In some situations, a proper integration of semantics extracted from nontextual resources (especially images) with the semantics learned from the text which accompanies them is very promising.
- Information integration: When combining information from different sources, aspects of provenance play a crucial role, since the quality and reliability of the sources may vary to a large extent. Typically, some information might be vague or uncertain, or only be valid in some periods of time. As a consequence, one has to develop methods that can deal with these different kinds of information that provide heuristics to combine information in these settings and are able to reason in these heterogeneous settings. In essence, methods from nonstandard logics, Bayesian networks and the like have to be combined with the more standard approaches that have been developed in recent years, like, for example, OWL.
- Advanced ontology mapping: Today's ontology languages do not include any provision for representing and reasoning with uncertain information. However, typical future application scenarios will lead to ontologies that are composed of concepts that are to some extent valid in a domain, relationships that hold to some degree of certainty, and rules that apply only in some cases. That is, we have to deal with ontologies that go beyond the area of standard logics. As such, approaches for ontology alignment or merging have to be extended to cover these challenges.

Whereas individual (non)logical approaches exist to address these aspects, one lacks a coherent framework to handle these challenges in an integrated way. How to provide methods that still scale up or how to design the interaction with the users in such complex scenarios, is still an open research issue.

14.5. SOCIAL SEMANTIC DESKTOP

In a complex and interconnected world, individuals face an ever-increasing information flood. They have a strong need for support in automatic structuring of their personal information space and maintaining fruitful communication and exchange in social networks within and across organizational boundaries. The realization of such a *Social Semantic Desktop* poses several challenges:

- Personal perspective of knowledge: Since more and more individual knowledge work is reflected in the information objects and file structures within the personal desktop, new techniques and methods are required to extract, structure, and manage such knowledge. In particular, the support to annotate and link arbitrary information on the local desktop, across different media types, file formats, and applications is needed as well as means for the quick, easy, and unintrusive articulation of human thoughts. The next step is to integrate content creation and processing with the users' way of structuring and performing their work.
- Knowledge work perspective: Knowledge work is typically task oriented. Therefore, dynamic task modeling is needed to provide the basis for context-sensitive annotation, storage, retrieval, proactive delivery, and sharing of information objects. Process-embedded usage of the support tools, taking into account personal experiences, will result in a comprehensive and goal-oriented information support for the individual knowledge worker.
- Social perspective: Individual knowledge work in practice never stands alone, but is integrated into communication, collaboration, and exchange between individuals connected via social networks. Keeping in mind privacy and access rights, each personal desktop can be considered as a peer in a comprehensive peer-to-peer network which facilitates distributed search and storage. More powerful methods and tools which transform a set of hitherto unrelated personal work spaces into an effective environment for collaborative knowledge creation and exchange across boundaries are needed. Furthermore, they will offer the user the means to link and exploit other people's knowledge, to comment and annotate other people's articulations and collaborate on shared knowledge bases.

The social semantic desktop realizes the vision of the so-called high performance workspace that will empower a knowledge worker in critical decision-making processes. However, meeting specific needs of knowledge workers in a particular context in order to attract their attention (so-called attention management) is a new, very challenging issue.

14.6. APPLICATIONS

Ontologies are a very promising technology for a variety of application areas, as discussed in numerous cases studies in the chapters of this book. Some are still to come: intelligent environments (contextually-appropriate personalised information spaces), personal knowledge networking (see the discussion on the Social Semantic Desktop above), and business performance management (i.e., near-real-time semantic information integration of critical business performance indicators to improve the effectiveness of business operations and to enable business innovations), to name but a few.

Moreover, in a light-weight form, ontologies are already used for structuring data in some popular web applications (e.g., flickr) or even in an industrial environment (e.g., in the form of corporate taxonomies). There are two main challenges for the wide industrial uptake of heavy-weight ontologies: (i) their formal nature that could decrease the readiness for a large-scale industrial adoption, and (ii) the lack of practical evidence (e.g., large-scale success stories) that clearly show the added value of applying semantic technologies. However, by having first commercial products on the market, for example, for knowledge management or information integration, there is now a promising opportunity to come up with more well analyzed application scenarios that show how ontology-based applications provide a real return of investment.

Looking beyond applications in knowledge and information management, work on standards for Semantic Web Services has already begun at the W3C. Semantic Web Services (SWS) aim to use semantic descriptions of services to enable automatic discovery, composition, invocation, and monitoring of web services and have the potential to impact significantly on IT integration costs and on the speed and flexibility with which systems can be (re)configured to meet changing requirements. SWS are discussed in detail in Chapters 10 and 13 of this volume. Beyond this, semantic technology will be applied to the Grid and in the area of pervasive computing. In the Grid context, the vision is that information, computing resources, and services are described semantically using languages such as RDF and OWL. Analogously to Semantic Web Services, this makes it easier for resources to be discovered and joined up automatically, which helps bring resources together to create the infrastructure to support virtual organizations. Pervasive computing envisions a world in which computational devices are ubiquitous in the environment and are always connected to the network. In the pervasive computing vision, computers and other network devices will seamlessly integrate into the life of users, providing them with services and information in an 'always on,' context sensitive fashion. Semantic technology can make a significant contribution by supporting scalable interoperability and context reasoning in such systems.

Finally, new types of application scenarios that exploit the sharing of information and on-the-fly cooperation between applications, require new trust and incentive models. These are key issues for both public and private sector organizations and must supplement the advancement of other semantic technologies in order to realize the full potential of the Semantic Web.

Index

Semantic Web Technologies: Trends and Research in Ontology-based Systems
John Davies, Rudi Studer, Paul Warren © 2006 John Wiley & Sons, Ltd